Astrophysics

ADVANCED
PHYSICS
READERS

Christopher
Bishop

JOHN
MURRAY

Other titles in this series:
Particle Physics   0 7195 8589 9

First published in 2000
by John Murray (Publishers) Ltd
50 Albemarle Street
London W1X 4BD

Layouts by Eric Drewery
Illustrations by Art Construction
Cover design by John Townson/Creation

Typeset in 11.5/13pt Goudy by Wearset, Boldon, Tyne and Wear
Printed and bound in Italy by G. Canale

A catalogue entry for this title is available from the British Library

ISBN 0 7195 8590 2

Contents

Introduction

This book has been written principally for A-level Physics students who are studying Astrophysics and Cosmology in their Physics syllabuses.

Much of the content assumes an understanding of 'core physics' concepts, such as basic mechanics, heat and properties of matter, waves and vibrations, radioactivity and atomic structure. The mathematical skills required are those of GCSE Higher level with the addition of some further concepts, which are explained in the text. At the end of each chapter is a summary of the main points together with a number of self-assessment questions which allow students to test their understanding. Some of the questions are more searching, but the student will learn valuable lessons by attempting all of them. Answers with outline solutions are provided at the back. Recent past paper examination questions from popular examining boards are also included. While not strictly examinable at A-level, I have attempted to introduce some more advanced ideas both quantitatively and qualitatively, such as free-fall collapse, stellar models and the formation of stellar spectra. It is hoped that these may be of value to students contemplating Astrophysics at degree level. In addition, I have also tried to give the student an understanding of how astrophysical data are obtained, together with an appreciation of its limitations.

Astrophysics is an exciting and dynamic subject, and a book of this size cannot do full justice to the huge range of phenomena and discoveries made in recent years by both ground and orbiting observatories. For this reason an associated website at *http://www.ph.surrey.ac.uk/astrophysics/index.html* has been created. This is maintained and run by the Physics Department of the University of Surrey; as well as finding abridged versions of the text, students will be able to 'hot link' to sites of current astrophysical interest and send in their own questions. This site is under continual development and it is hoped that, together with this book, student and teacher alike will find this to be a valuable resource for their understanding of the subject.

Finally, I hope that *Astrophysics* will give you, the reader, an introduction to the awesomeness and wonder of the universe as so aptly captured by this extract from Wordsworth's poem *Lines Composed a Few Miles Above Tintern Abbey*:

> *For I have learn'd*
> *To look on Nature, not as in the hour*
> *Of thoughtless youth; but hearing oftentimes*
> *The still sad music of humanity*
> *Nor harsh nor grating, though of ample power*
> *To chasten and subdue. And I have felt*
> *A presence that disturbs me with the joy*
> *Of elevated thoughts; a sense sublime*
> *Of something far more deeply interfused,*
> *Whose dwelling place is the light of setting suns,*
> *And the round ocean, and the living air,*

And the blue sky, and in the mind of man;
A motion and a spirit that impels
All thinking things, all objects of all thought,
And rolls through all things. Therefore I am still
A lover of the meadows, and the woods
and mountains; and all that we behold
From this green earth.

Acknowledgements

Writing a book is a rewarding task that requires both mental stamina and attention to detail. Many people have been involved in producing *Astrophysics* and I am indebted to them all. I would particularly like to express my thanks to the following: my editorial team at John Murray – Katie Mackenzie Stuart (Science Publisher), for inviting me to write it and for all her encouragement and guidance, Julie Jones (Science Editor), whose management and production skills ensured that this book saw the light of day, Abigail Woodman, whose ability to research and obtain the most obscure of astronomical images knows no bounds, and Jane Roth, whose skilful editing greatly enhanced its readability; Keith Fuller of the Guildford High School for Girls, who read through the manuscript and made many valuable comments and suggestions; my colleagues – Jim Al-Khalili, for discussions on the nature of Black Holes and for suggestions on the relevant sections, and David Wonnacott, who checked many of the questions and answers for accuracy and consistency. I would also like to thank Professor Eoin O'Reilly, Head of Department of Physics at the University of Surrey, for his support. Last, but not least, thanks to Sally for putting up with an unsociable spouse despite carrying my daughter Charis, who was born while this book was being written.

For my son and future astronaut, Ciaran.

Examination questions

Exam questions have been reproduced with kind permission from the following examination boards. The numerical answers provided have been written by the author, and the examination boards bear no responsibility for their accuracy.

Edexcel (formerly ULEAC)
OCR (formerly UCLES)

Source acknowledgements

The following are sources from which artwork and data has been adapted:

Figure 1.8a, p.16 A.E. Roy and D. Clarke *Astronomy: Structure of the Universe*, Adam Hilger, 1989
Figure 2.2, p.25 R.C. Smith *Observational Astrophysics*, Cambridge University Press, 1995
Table 3.2, p.54 *Norton's 2000.0 Star Atlas and Reference Handbook* 18th edition, © Longman Group UK Limited 1989. Reprinted by permission of Pearson Education Limited
Figure 3.3, p.60 S. Inglis *Planets, Stars and Galaxies* 4th edition, John Wiley & Sons Inc, 1976

Figure 3.4, p.61 Cole, Franklyn and Wellington *Fundamental Astronomy: Solar System and Beyond*. Copyright © 1974 John Wiley & Sons Inc. Reprinted by permission of John Wiley & Sons Inc
Figure 3.12, p.73 C.R. Kitchin *Astrophysical Techniques*, Adam Hilger, 1991
Figure 5.5b, p.110 M. Zeilik and J. Gaustad *Astronomy: The Cosmic Perspective* 2nd edition, John Wiley & Sons Inc, 1990
Figure 5.4, p.107; Figure 7.07, p.139; Figure 8.2, p.151; Figure 8.3, p.153; Figure 9.5, p.163; Figure 10.7, p.201; Figure 10.9, p.208 UNIVERSE by Kaufman © 1985, 1988, 1991, 1994, 1999 by W.H. Freeman and Company. Used with permission
Figure 10.3, p.191 J.A. Wheeler *Journey into Gravity and Space Time*, Scientific American Library, 1990

Photo credits

Thanks are due to the following for permission to reproduce copyright photographs:

Cover Mount Stromlo & Sliding Spring Observatories/Science Photo Library; **p.2** Mary Evans Picture Library; **p.5** Linda Whitnall; **p.6** Jack Finch/Science Photo Library; **p.7** Adrian Meredith Photography; **p.8** C.A. Grady (National Optical Astronomy Observatories, NASA Goddard Space Flight Center), B. Woodgate (NASA Goddard Space Flight Center), F. Bruhweiler and A. Boggess (Catholic University of America), P. Plait and D. Lindler (ACC, Inc., Goddard Space Flight Center), M. Clampin (Space Telescope Science Institute), and NASA. The ground-based image is courtesy of P. Kalas (Space Telescope Science Institute); **p.10** Observatoire Pic du Midi, France; **p.11** *t* Hale Observatories/Science Photo Library, *b* NASA/JPL image courtesy of Armagh Planetarium; **p.13** NASA/JPL image courtesy of Armagh Planetarium; **p.16** By kind permission/Courtesy of The London Planetarium; **p.17** Copyright © UK ATC, Royal Observatory, Edinburgh and Anglo-Australian Observatory; **p.18** Harvard-Smithsonian Center for Astrophysics; **p.19** NASA STSCI-PRC95-04 image courtesy of Armagh Planetarium; **p.68** Richard Smith/Galaxy Picture Library; **p.69** NASA/Science Photo Library; **p.70** NASA STSCI-79-72 image courtesy of Armagh Planetarium; **p.74** John Walsh/Science Photo Library; **p.76** Galaxy Picture Library; **p.81** *l & r* Nial Tanvir Institute of Astronomy, Cambridge; **p.86** Courtesy of the Physics Department, University of Surrey; **pp.89, 90** California Institute of Technology; **pp.91, 92, 93** *all* Yerkes Observatory; **p.114** NASA STSCI image courtesy of Armagh Planetarium; **p.116** © Anglo-Australian Observatory; **p.118** Rev Ronald Royer/Science Photo Library; **p.119** Palomar Observatory/Caltech; **p.120** Copyright © UK ATC, Royal Observatory, Edinburgh and Anglo-Australian Observatory; **p.121** NASA/ Science Photo Library; **p.142** Galaxy Picture Library; **p.155** NASA STSCI-PRC95-01 image courtesy of Armagh Planetarium; **p.160** *t* Yerkes Observatory, *b* UCO/Lick Observatory image; **p.164** ©Anglo-Australian

Observatory; **p.165** NASA STSCI-PR94-22 image courtesy of Armagh Planetarium; **p.166** Palomar Observatory/Caltech; **p.169** University of Manchester, Jodrell Bank Observatory; **p.178** Courtesy J. Kristian; **p.194** Palomar Observatory/Caltech; **p.204** Lawrence Berkeley Laboratory/Science Photo Library; **p.211** NASA/Science Photo Library.

t = top, *b* = bottom, *l* = left, *r* = right

The publishers have made every effort to contact copyright holders. If any have been overlooked, they will be pleased to make the necessary arrangements at the earliest opportunity.

1 The universe at large

'The introduction begins like this:
"Space," it says "is big. Really big. You just won't believe how vastly hugely
mindbogglingly big it is. I mean you may think it's a long way down the road to the chemist,
but that's just peanuts to space. Listen . . ."'

from *The Hitch Hiker's Guide to the Galaxy* by Douglas Adams

What is astrophysics?

Since ancient times people have looked up at the stars and wondered about their place in the universe. The view of the stars then was not obscured as it is today by modern widespread use of street lighting and other forms of light pollution, which rob us of the brilliance and splendour of the night sky.

Early people regarded the Sun as the giver of life and worshipped it (as they did the Moon). The builders of Stonehenge and other megalithic monuments observed the rising of the Sun and the phases of the Moon and planted their crops in accordance with the celestial calendar. Some of the ancients thought they saw the outlines of men and beasts in the heavens and named them after heroes, animals and beings from mythology, giving us the names of the constellations we know today.

We learn from their writings that around 3000BC the Babylonians believed that the positions and motions of all celestial bodies influenced the fate of human lives. This belief gave rise to **astrology**. While having no scientific basis, astrology was significant in stimulating the study and recording of celestial bodies including the development of observational techniques that became useful in **astronomy**, the oldest of the exact sciences.

Astronomy is a discipline primarily concerned with the systematic observation and recording of celestial objects including phenomena such as solar and lunar eclipses, comets, novas and phases of the Moon. Foremost among the ancient astronomers was Hipparchus, who was born in Nicaea (now Iznik) in northwest Turkey in about 190BC and is considered to be the founder of astronomy as a scientific discipline. Hipparchus calculated the distance of the Moon from the Earth using the method of **parallax** (see Chapter 3). He also produced the first accurate star map and devised a system of 'star brightness' based on a magnitude scale which, although later refined, is still used today.

The invention of the telescope in early 17th century revolutionised astronomy and many new discoveries were made. Running parallel with this development was Sir Isaac Newton's publication of his Universal Law of Gravitation which, together with his investigations into the nature of light, provided the beginnings of a union between astronomy and physics.

Today, **astrophysics** is the study of the physical properties and composition of celestial objects using the known laws of physics. Most of the information about the physical nature of the astronomical universe comes to us by means of electromagnetic radiation which celestial objects emit at various wavelengths. By analysing the intensity and form of the spectrum of the radiation, astrophysicists can learn much about the physical conditions inside stars, in interstellar space, the nature of galaxies and can even postulate what the early universe was like.

Figure 1.1 shows a portrait of German physicist Gustav Robert Kirchhoff (1824–87) who is generally acknowledged as the 'founding father' of astrophysics. Kirchhoff developed the first **spectroscope**, which enabled spectral lines to be observed when chemical elements were heated to high temperatures. It soon became apparent to

Figure 1.1 Gustav Robert Kirchhoff (1824–87) whose three rules governing the formation of different types of spectral lines laid the foundations of modern astrophysics

Kirchhoff that each element had its own unique set of spectral lines. For example, hot sodium vapour produces a characteristic double yellow line. Through the use of a spectroscope, the chemical composition of a star can be determined as each element leaves a 'signature' in a characteristic pattern of lines.

Kirchhoff was the first to analyse light from the Sun using a spectroscope, and discovered half a dozen spectral line patterns corresponding to elements observed spectroscopically on Earth. From this, he deduced that these elements must also exist in the Sun. He also formulated three rules of spectroscopy that are of fundamental importance in astrophysics and we will have more to say about them in Chapter 4.

Nowadays many scientists use the words astronomy and astrophysics more or less synonymously but it should be noted that astronomy is primarily an *observational* science concerned with the motions and distributions of celestial objects, while astrophysics is the study of the *physical properties* of celestial objects and the interaction of matter and energy within them. Loosely stated,

$$\text{astronomy} + \text{physics} = \text{astrophysics}$$

However, before we look at astrophysics in detail, let us go on a short guided tour of the universe and summarise the principal kinds of objects it contains.

The Earth in space

The Earth on which we live is an example of a **planet**. The shape of the Earth is very nearly spherical with an equatorial diameter of about 12 756 km and a mass of 6×10^{24} kg. Approximately 70% of the Earth's surface is covered in water and the average density is about 5500 kg m^{-3}. It rotates once every 24 hours and orbits the Sun with an average speed of 29.8 km s^{-1} at a mean distance of 1.49×10^8 km and with an orbital period of just over 365 days. The shape of the orbit is nearly circular and the Earth's axis of rotation is tilted by 23° 27' normal to the plane of the orbit, giving rise to our seasons of spring, summer, autumn and winter.

The Earth possesses an atmosphere composed of nitrogen (78%), oxygen (21%) and argon (0.9%) with the remainder being trace gases of other elements. The thin layer of atmosphere extends to an altitude of some 1000 km and has a complex weather system driven by thermal air currents powered by solar radiation. It is only transparent to a certain range of wavelengths and the sky appears blue due to the scattering of sunlight by atmospheric dust particles.

The interior of the Earth is thought to be composed of a solid **inner core** of iron and nickel, a molten **outer core** of liquid iron-nickel and a **mantle** made up of rock comprising iron and magnesium combined with silicon and oxygen. On top of this is the solid **crust** which is the rocky surface that we live on. The temperature in the outer core is about 2600 K and the source of heat is believed to come from trapped radioactive materials such as isotopes of uranium and thorium converting nuclear energy to thermal energy.

The Earth also possesses a magnetic field of average strength 5×10^{-5} T, the source of which is thought to be due to the action of the fluid part of the core behaving like an electrical generator or **dynamo**. As the liquid core (which is metallic) moves due to the Earth's rotation, it swirls around generating magnetic fields which permeate the interior and escape through the crust. This dynamo model suggests that any planet which has a strong external magnetic field probably has a fluid conducting core which is in motion.

Table 1.1 shows the ten most abundant elements found in the Earth. One of the tasks of astrophysicists is to explain how these (and other elements in the periodic table) were created in the abundances we see them today, not only in the Earth, but also in the universe as a whole.

Table 1.1 The ten most abundant elements in the Earth

Element	Average % by mass	Element	Average % by mass
Iron (Fe)	34.6	Sulphur (S)	1.9
Oxygen (O)	29.5	Calcium (Ca)	1.1
Silicon (Si)	15.2	Aluminium (Al)	1.1
Magnesium (Mg)	12.7	Sodium (Na)	0.57
Nickel (Ni)	2.4	Chromium (Cr)	0.26

The age of the Earth can be estimated from radioactive dating of rocks. Using the decay of rubidium ($^{87}_{37}Rb$) into strontium ($^{87}_{38}Sr$), geologists estimate the age of the Earth to be approximately 4.0×10^9 years.

The temperature range on Earth varies from a maximum of 60 °C to a minimum of −90 °C. Life on our planet depends very critically on maintaining a stable orbit and distance from the Sun. Too close and the temperature would be too hot to sustain life, and if its distance from the Sun were any greater then we would all freeze to death! This implies that the Earth must have been in a stable orbit for at least as long as life has existed on it.

The Solar System

The Earth inhabits a system of nine planets including a belt of minor planets or **asteroids** (bodies less than 1000 km in diameter), which are all associated by the fact that they all move in elliptical orbits around the Sun due to its gravitational attraction. The Sun and its family of planets is called the **Solar System**.

The orbits of all the planets are nearly circular and all lie more or less in the same plane. In order outwards from the Sun, the nine major planets are: Mercury, Venus, Earth, Mars, Jupiter, Saturn, Uranus, Neptune and Pluto. With the exception of Mercury and Venus, all have smaller satellites called **moons** orbiting them. Jupiter has 16 known satellites, four of which are easily visible using binoculars.

It is an interesting coincidence that the Sun and the Earth's Moon have nearly the same angular diameter in the sky. Occasionally the Moon will pass between the Sun and the Earth, covering the Sun partially or totally. This spectacular event is called a **solar eclipse** and happens several times a century. During a total eclipse, the bright sunlight is extinguished and the outer atmosphere of the Sun called the **solar corona**, extending for millions of kilometres into space, becomes visible.

Astronomers divide the planets into two groups: **terrestrial** and **Jovian**. The terrestrial planets are Mercury, Venus, Earth and Mars and consist largely of iron and silicate rocks. The Jovian, or Jupiter-like, planets are Jupiter, Saturn, Uranus and Neptune and are composed largely of hydrogen and helium. The terrestrial planets lie closer to the Sun whereas the Jovian planets are larger, more massive, have stronger magnetic fields and rotate more rapidly on their axes. In addition, unmanned space probes have shown that all the Jovian planets have ring systems with moons orbiting them. The outermost planet, Pluto, is small and somewhat unusual: its orbit has more in common with asteroids than the other planets, and its surface is thought to be made of methane ice. It does not fit very well into either category and astronomers have speculated that it might once have been a moon of Neptune and not a planet at all.

Unlike stars, planets do not generate light of their own but shine by *reflected light*. We observe the planets of the Solar System and their moons because sunlight is reflecting off them even though the Sun is well below our horizon in the night sky.

Occasionally **comets** moving in highly elliptical orbits around the Sun may be seen from Earth. These consist of a nucleus made up of small particles encased in

Figure 1.2 Comet *Hale–Bopp* over Avebury Stone Circle, Wiltshire, in 1997. Photographed by Linda Whitnall

frozen gases. As a comet nears the Sun, evaporation of the gases occurs and an **ionised tail** develops. The tail always points away from the Sun due to the **solar wind** – a stream of charged sub-atomic particles (protons and electrons) that recede from the Sun at high velocities. The solar wind carries a magnetic field which drags ions in the comet's tail along with it, making it behave rather like a wind sock! The origin of comets is still a matter of some speculation, but it is thought that a cometary cloud called the *Oort Cloud* (postulated by the Dutch astronomer Jan H. Oort (1900–92)), may exist far beyond the orbit of Pluto.

Most of the comets in the *Oort Cloud* never get near the Sun, but sometimes the action of a passing star may push one of them into an orbit that causes it to pass close enough to enter the inner Solar System. A spectacular comet *Hale–Bopp* recently passed our way and you may have seen it (Figure 1.2).

Meteoroids are lumps of rock that the Earth encounters in space as it travels in its orbit. If one enters the atmosphere it heats up as a bright streak of light called a **meteor**. Most meteors burn up before they reach the ground but those that reach the surface of the Earth are called **meteorites**. Meteorites have masses ranging from a few grams to several tonnes and there is some evidence that very large meteorites have hit the Earth in the distant past causing great environmental and climatic

Figure 1.3 (a) The aurora borealis. This spectacular display of coloured light streamers is caused by an increase in the solar wind

changes. By analysing the chemical composition of meteorites, we can discover clues about their origins and the abundances of the chemical elements in the Solar System.

As well as comets, meteoroids and asteroids, interplanetary space also contains dust (see Chapter 6), and sparse amounts of hydrogen gas, protons and electrons. These and other particles ejected by the Sun in the solar wind dominate interplanetary space even at distances well beyond the orbit of Pluto.

The intensity of the solar radiation depends on how active the Sun's surface is. When a **solar flare** (see page 12) occurs a dramatic increase in the solar wind takes place. On Earth, this can cause atmospheric disturbances such as the **aurora borealis** or 'northern lights' (Figure 1.3a) and high-flying aircraft such as Concorde (Figure 1.3b) carry alarms to warn of the increased radiation.

The Solar System is thought to have been formed from a rotating primordial **solar nebula** or gas cloud. A rotating body has **angular momentum** (see Box 9.3, page 170). An effect of the conservation of angular momentum was to flatten the rotating gas cloud into a 'plate' (rather like wet clay on a potter's wheel). Local aggregations of matter formed 'clumps' of gas from which massive bodies formed and eventually condensed to give the Sun and planets as we know them today.

Figure 1.3 (b) The increase in solar radiation can cause alarms to go off in high-flying aircraft such as Concorde, which must then reduce altitude to protect passengers and crew from excess exposure

It is almost certain that stars other than the Sun have planetary systems. Figure 1.4 shows a picture of a developing star A B *Aurigae* with what appears to be a disk of matter containing clumps of dust and gas surrounding it. Are we seeing the formation of planets in this image?

Figure 1.4 Hubble Space Telescope image of the developing star *A B Aurigae*. The image reveals a swirling disk of dust and gas surrounding the star at the centre. Clumps of matter can be seen, which may be the seeds of planet formation. *A B Aurigae* lies in the constellation *Auriga*

The Sun and other stars

On a clear night you can probably see many thousands of stars, each appearing as bright points of light. Some stars appear brighter than others and, if you look carefully, they even show colours. If we look through a telescope then we see myriad other stars which are too faint to be seen with the naked eye.

What is a star?

Stars are intensely hot, gaseous, spherical bodies which produce energy by means of **thermonuclear reactions** converting hydrogen into helium (see Chapter 7). The basic structure of a star can be considered as a series of layers, each having its own energy transfer process. At the centre is a core which is burning by nuclear reactions. This energy is transported away first by radiation, then by convection, and when it reaches the surface of the star it escapes into space as electromagnetic radiation (Figure 1.5).

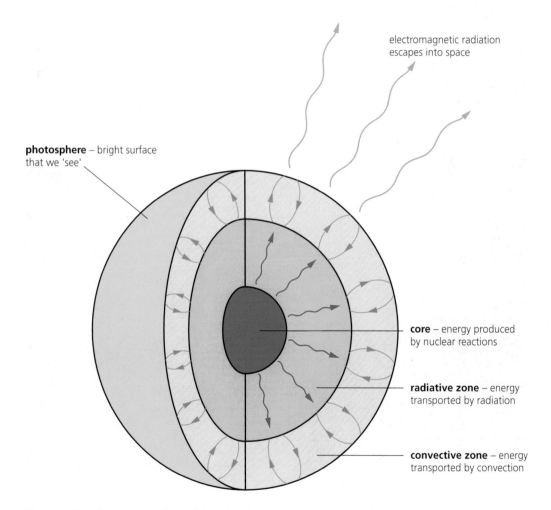

electromagnetic radiation escapes into space

photosphere – bright surface that we 'see'

core – energy produced by nuclear reactions

radiative zone – energy transported by radiation

convective zone – energy transported by convection

Figure 1.5 Basic structure of a star showing the principal energy transfer processes

Most stars share common features, and the star that we know the most about is the Sun. The Sun has a mass of 2×10^{30} kg and a diameter of about 1.4×10^6 km. Its average density is 1.4×10^3 kg m^{-3} and it has a surface temperature of about 6000 K. It rotates on its axis with a period of 25 days and possesses a magnetic field of strength 0.1 to 0.2 mT at the surface.

The bright surface of a star such as the Sun is called the **photosphere**. This is the visible surface of the Sun that we can see from the Earth, and is covered with a pattern of bright grainy markings called **granulations**. These granulation patterns continually change as hot gases flow out from the interior to the surface (Figure 1.6a).

Also on the photosphere can be seen dark patches called **sunspots** which vary in size from a few kilometres to many times the size of the Earth. A typical sunspot consists of a dark nucleus called the **umbra** surrounded by a lighter area called the **penumbra** (Figure 1.6b). Sunspot temperatures are some 2000 K lower than the average temperature of the photosphere, which is why they appear dark in contrast. They tend to occur in groups and are associated with regions of high magnetic field intensity with values reaching as high as 0.4 T.

Sunspots always come in pairs. The magnetic fields in any one pair always have opposite polarities. Magnetic field lines emerge from the interior of the Sun through one member, loop through the solar atmosphere and re-enter the photosphere through the other spot (Figure 1.6c). Observations of the Sun over many centuries

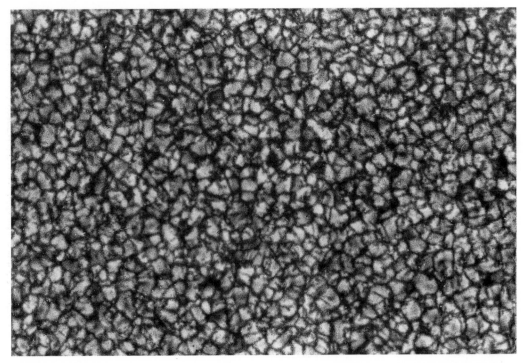

Figure 1.6 (a) Granulations on the photosphere of the Sun. These are areas where hot gases flow out from the convective zone to the surface

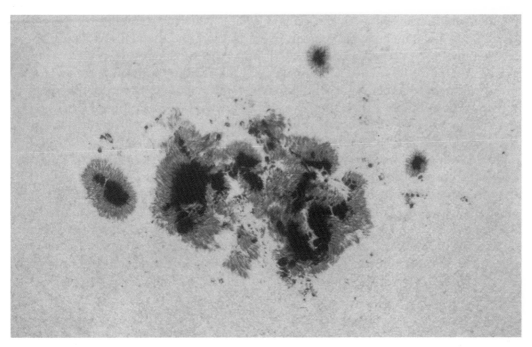

Figure 1.6 (b) Sunspots on the photosphere. These are cooler areas of the surface associated with regions of high magnetic field intensity

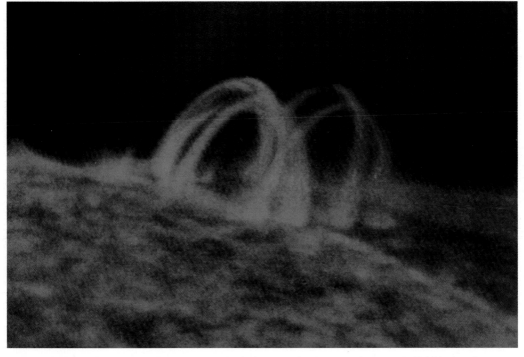

Figure 1.6 (c) This picture, taken in ultraviolet light, reveals patterns of magnetic field loops above solar active regions

have shown that sunspot activity occurs in cycles. The average number of spots builds up to a maximum approximately every 11 years and then falls off almost to zero before the cycle begins again. In addition to this **sunspot cycle**, there is a longer **solar cycle** lasting 22 years. During a sunspot cycle, leading pairs of spots of a group in the northern solar hemisphere all have the same magnetic polarity, while spots in the southern hemisphere have the opposite polarity. These polarities reverse their signs after each 11-year cycle, so, when the Sun's magnetism is taken into account the full cycle takes 22 years.

The Sun has an atmosphere of ionised gases that extends to about 2500 km, called the **chromosphere**. The temperature of the chromosphere is much higher than the photosphere with values ranging up to 28 000 K. At the edge of the chromosphere is the **transition region** where the temperature increases sharply into the solar **corona**, a region of rarefied ionised gas called a **plasma**, with temperatures up to a few million kelvin. The corona extends hundreds of thousands of kilometres into space and is not normally visible, but it can be seen during a total eclipse of the Sun. Table 1.2 summarises how the temperature of the Sun's atmosphere varies above its surface.

Table 1.2 Variation of temperature with height above the surface of the Sun

Region	Height above the surface/km	Temperature/K
Photosphere	0–320	6500–4500
Chromosphere	320–1990	4500–28 000
Corona	~7 × 10^5	1.8 × 10^6

Flares are violent discharges of energy associated with sunspots. A large flare can release some 10^{25} J of energy lasting from 15–20 minutes to as long as a few hours. Flares shoot out energetic protons and electrons into space but the majority of the radiation is in the form of X-rays and ultraviolet radiation. Travelling at the speed of light, this arrives at the Earth in about eight minutes, ionising the upper atmosphere and causing radio communications to be disrupted. The slower-moving protons and electrons arrive about a day later and are trapped in the Earth's magnetic field where the electrons collide with atmospheric atoms and excite them to higher energies. As the atoms de-excite they emit light, giving rise to the aurora borealis mentioned earlier.

Prominences are huge eruptions of hydrogen gas above the photosphere attaining a height of many thousands of kilometres (Figure 1.7). Like flares, they are usually associated with sunspots and can loop and change shape as they surge up into the corona. Unlike flares they can last for many weeks as the solar material 'rains down' onto the photosphere. Flares and prominences create 'thunderstorms' in the solar wind and the excess radiation produced by them can be a hazard to astronauts and satellites in space.

Figure 1.7 A huge solar prominence. This picture, taken in ultraviolet light, was obtained during the *Skylab 3* mission and shows the size of the prominence in relation to the white dot, which represents the size of the Earth

Different types of star

As well as stars like our Sun, there are many types of star with a range of both size and temperature. *Cool* stars are stars with surface temperatures between 1600 K and 2500 K which radiate with a deep red colour, while *hot* stars have surface temperatures as high as 35 000 K and glow bluish-white. Stars range in size from **white dwarfs** with dimensions similar to that of the Earth to **supergiants** with diameters several hundred times that of the Sun. They can range in mass from below 0.1 to 100 solar masses. We will learn more about the different stars and how they live out their lives in Chapter 8. There also exist exotic stars such as neutron stars and black holes. These are really ex-stars which no longer shine and we will discuss them separately in Chapter 9.

The distance between the stars is very great. Astronomers use a unit called the **light year** (ly) to measure distances in interstellar space. A light year is the distance travelled by a beam of light in one year. Since light travels at 3×10^8 m s^{-1} and there are $365 \times 24 \times 60 \times 60 = 3.15 \times 10^7$ s in a year, then 1 light year $= (3 \times 10^8$ m s$^{-1}) \times (3.15 \times 10^7$ s$) = 9.46 \times 10^{12}$ km. Another unit used is the **parsec** (pc) which is equal to 3.1×10^{13} km. (The light year and parsec are defined more precisely in Chapter 3.) The nearest star to the Earth (other than the Sun) is the *Alpha Centauri* star system at just over four light years away.

Variable stars

An important class of stars is that of the variable stars. These are stars that vary in brightness and are classified into two kinds: **eclipsing variables** and **intrinisic variables**.

As many as 50% of stars exist as binary pairs. These **binary stars** are double stars that consist of two components orbiting each other. Some stars have even been resolved in telescopes to consist of three or more stars gravitationally bound together. As one of these stars eclipses another, the apparent brightness of the composite object changes in a regular fashion by which a graph of brightness versus time called a **light curve** can be drawn.

Intrinsic variables are stars in which the variation in brightness is due to *physical changes* in the internal structure of the star and not because of any orbiting companion. An important type of intrinsic variable star is a **Cepheid variable**. These variable stars vary in brightness due to pulsations of their photospheres and atmospheres. Later on we will see that they are very useful as **distance indicators**, i.e. they can be used as a 'yardstick' for determining astronomical distances. Binary star systems also enable us to estimate a star's mass using Kepler's Laws (see page 38) by observing the changes in their angular separation and period.

Novas and supernovas

Occasionally, we may be treated to the spectacle of a **nova** or a **supernova**. These are quite literally exploding stars in which a sudden release of energy within the star increases its brightness by hundreds and thousands of times over a period of a few days. A nova is believed to occur in a binary system where matter from one star falls onto the other, creating violent detonations on its surface. Supernovas are even more catastrophic, occurring when a massive star, having used up its fuel, rapidly collapses in on itself producing a shock wave of enormous energy – as much as 10^{44} J. This is roughly the total amount of energy that our Sun will produce in its lifetime!

Supernovas mark the endpoints of massive stars. In a supernova, matter from the interior of the star is blasted into space at tremendous speeds. Supernova explosions make many of the heavier elements in the periodic table and these provide the 'building blocks' from which other stars and planets are made. Everything contains elements that were made at some distant time in the past during a supernova event, and you and I are all in a very real sense made of supernova stardust!

The interstellar medium

The space between the stars is known as the **interstellar medium (ISM)**. The ISM is not empty but contains dust, gas and large clouds of atomic and molecular hydrogen including a variety of organic and inorganic molecules. Where does all this material come from?

Astrophysicists believe that when the universe was formed matter and energy were created in a primordial fireball called the **Big Bang**. A few thousand years after the Big Bang, the universe had cooled to a temperature in which the simplest atoms such as hydrogen and helium were formed plus smaller amounts of other light elements such as lithium and boron. As time went on this accumulation of matter condensed under gravity into giant clouds which eventually formed the stars and galaxies.

The ISM is not static but a dynamic system, as it is the 'stuff' from which stars are made. As stars form they withdraw mass from the ISM and during their lifetimes form many heavier elements in their interiors. As they shine, a small amount of their mass is returned to the ISM as matter blows off their surface in the form of 'stellar winds' but at the end of their lives they return most of their mass (now chemically enriched by nuclear reactions) either by slow shedding of their outer layers, or by explosive events such as novas and supernovas. The ISM is therefore a kind of interstellar ecosystem in which stars are born, live and die. In Chapter 6 we will explore in more detail how this works.

Since the ISM is mainly gaseous, we can study it by observing the spectral lines it emits or absorbs and from this we can determine its density, composition, temperature and mass distribution. With this information astrophysicists can construct models of star formation which explain how stars are born out of gravitational and nuclear processes and then magnificently start to shine.

Galaxies

Stars are not uniformly distributed throughout space but are gathered by gravitational attraction into huge collections of stars called **galaxies**. A galaxy may contain as many as 10^{11} stars and is classified according to its appearance, as spiral, elliptical or irregularly disk-shaped.

On a clear dark night away from any light pollution, when looking upwards you can see a faint band of light called the *Milky Way* stretching across the sky. What you are looking at is an edge-on view of the disk of our own galaxy. If you have keen eyesight then it is possible to trace the outline of the thickness of the disk and a low-powered telescope will reveal it to consist of millions of stars. Our own galaxy is a spiral galaxy approximately 30 000 pc across with an estimated mass of 10^{11}–10^{12} solar masses. It consists of a spherical central bulge about 5 kpc in radius with the disk about 1000 pc thick (Figure 1.8). The distance from our Solar System to the galactic nucleus is approximately 8500 pc.

Spread throughout the disk are huge 'dust lanes' containing atomic and molecular hydrogen. The fact that our galaxy has spiral arms suggests that it is rotating although not all parts of it are rotating at the same speed. The Sun takes about 2×10^8 years to complete one orbit about the galactic centre.

Orbiting at high galactic latitudes are **globular clusters**. These are spherical clusters of 10^4–10^5 stars which are quite old compared to average star lifetimes and define a galactic **halo**. About 100 globular clusters have been observed around the Milky Way and others have been discovered associated with other galaxies.

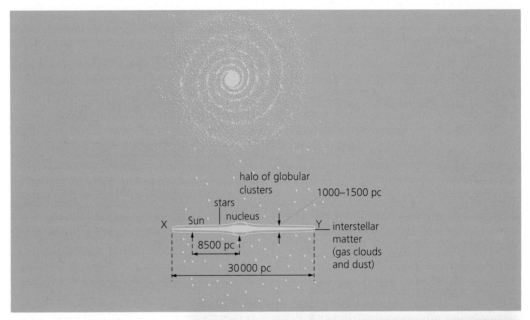

Figure 1.8 The *Milky Way* galaxy

(a, above) Our galaxy is a spiral type made up of a number of spiral arms. Looking along the direction X–Y from our Solar System we see a great many stars in the line of sight – this is the band of stars we call the Milky Way

(b, right) An artist's impression looking towards the centre of our galaxy

We are not able to see the centre of our galaxy due to obscuring clouds of interstellar gas and dust, but infrared and radio observations show that interstellar matter is rotating around the nucleus at very high velocities and is highly turbulent. Such motions could only be produced by a compact supermassive object with a high gravitational field and some astrophysicists have suggested that a **supermassive black hole** may exist at the centre of our galaxy. (Black holes are discussed in Chapter 9.)

Looking outwards from the Milky Way we see many more galaxies. About 77% are spiral types, 20% ellipticals and 3% irregular. We can observe their emissions at radio, infrared, visible and X-ray wavelengths, from which we can learn more about their structure.

Observations have shown that galaxies occur in **clusters**. The Milky Way galaxy belongs to one such cluster called the *Local Group*, containing at least 20 galaxies taking up a volume of space about 1 Mpc across (1 Mpc = 10^6 pc). Figure 1.9 shows the *Virgo Cluster*.

However, it doesn't stop with clusters! There exist 'clusters of clusters' of galaxies called **superclusters**, typically of size 75 Mpc across, composed of many smaller groups gravitationally bound together. An interesting feature of these superclusters is that the space between them is relatively empty with large 'voids' in which few or no galaxies exist and at this scale the architecture of the universe seems to be 'thread-like' or filamental.

Figure 1.9 The *Virgo Cluster*. This cluster contains as many as 2500 galaxies. Several spiral and elliptical galaxies can be seen here. The Virgo Cluster is about 20 Mpc from the Milky Way

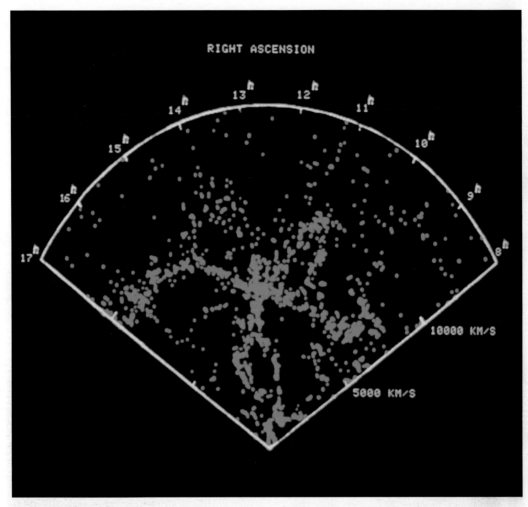

Figure 1.10 A wedge of galaxy distributions showing a filamental or thread-like appearance that resembles a 'stick man'

For some years the Harvard-Smithsonian Centre for Astrophysics has been conducting galaxy surveys in which the distances and locations of galaxies within a single 'wedge of space' (the apex of the wedge being our galaxy) are plotted. These wedges can be combined in a computer simulation to show the distribution of galaxies from different angles and perspectives. Figure 1.10 is one such projection showing a distribution of superclusters. This enormous structure is millions of light years across. It may in fact be much larger than this as the survey does not show it all. An important objective of modern astrophysics is to explain how large-scale galactic structures such as this could have formed and we will address this question again when we consider **cosmology** in Chapter 10.

Quasars

Some of the most distant objects known in the universe are termed Quasi-Stellar Objects (QSOs) or **quasars** (Figure 1.11). These appear to be point-like sources of light and radio waves that have very high **red shifts**. A red shift is an increase in the wavelength of the electromagnetic radiation received from an object as a result of the **Doppler effect** (see Chapter 2). The radiation is shifted towards the red end of the spectrum and indicates that the object is moving away from the observer. The greater the amount of shift the faster the object is receding. The high red shifts of quasars indicate that they are receding from us at speeds close to that of light. If these speeds of recession are due to the expansion of the universe then quasars are very far away indeed. Since the light from distant objects takes a finite time to reach us (due to the finite, constant speed of light), quasars must represent objects that were formed early on in the history of the universe and, since we can see them from such large distances, they must be extremely luminous – thousands of times more so than normal galaxies.

The energy emitted from a quasar (some 10^{40} W!) seems to come from a fairly small region of space at its centre, and some have been observed to vary their output on a time scale varying from days to years. Others can be seen ejecting jets of matter.

Astrophysicists think that these enigmatic objects may be an early form of galaxy. Quasars share some similarities with another class of objects called **active galaxies**. Active galaxies are closer to us but also have high luminosities (10^{37} W), rapid variability, peculiar shapes and jet-like emissions from their nucleus.

What is the energy source that drives quasars and active galaxies? As mentioned earlier, one idea is that there may be an extremely supermassive black hole at the centre of these objects which draws in passing stars and radiates huge amounts of energy in the process. A galaxy contains plenty of stars to feed the black hole's appetite and this model also accounts for the fact that the energy is emitted over a relatively small volume of space. Is a quasar then simply a young galaxy in an earlier stage of evolutionary development? This is one question that astrophysicists have yet to answer.

Figure 1.11 Quasars. These objects are very distant and are thought to be an early form of galaxy

How big is the universe?

The answer to this question depends in part on how far our observational instruments allow us to see into space, and what theories we have about how the universe came into existence. We shall discuss theories of the universe's origin or **cosmologies** in detail in Chapter 10. Our observational knowledge of the universe has advanced by leaps and bounds due to the advent of **space astronomy**.

Space astronomy is the use of rockets and satellites carrying telescopes and detectors high above the Earth where they can look out into space unhindered by the restricted view through the Earth's atmosphere. Space-based observatories can examine the universe at electromagnetic wavelengths not normally accessible to observers on Earth and transmit this information to ground stations where specialised signal- and image-processing techniques are used to extract the maximum amount of information from the raw data.

Table 1.3 shows a size and distance scale for the principal known objects in our universe. Our current observable limits in distance are about 300×10^6 ly although space observatories like the Hubble Space Telescope are continually pushing this boundary back. Because of the finite speed of light we view distant objects as they were many millions of years ago, so as we look out into space we are looking back in time. As the limits of observation are extended we should start to see more primitive structures, closer in time to the universe's creation.

Table 1.3 Sizes and distances in the universe

Diameter of atomic nucleus	10^{-14} m
Diameter of atom	10^{-10} m
Diameter of Earth	10^7 m
Diameter of Sun	10^9 m
Earth–Moon distance	4×10^8 m
Earth–Sun distance	1.5×10^{11} m
Diameter of Milky Way	10^{21} m
Distance to nearest galaxy	10^{22} m
Farthest galaxy seen	5×10^{25} m
Estimated number of stars in observable universe	$>10^{20}$

The story of the universe is still unfolding as new discoveries shed light on our understanding of space, time and matter. In this book, you will see how astrophysics helps us to understand more about the universe and enables us to make sense of the wide range of phenomena it contains as well as the physical processes that govern them. Much has still to be explained, but the endeavour to learn more about space goes on, driven by the ancient desire of humankind to know the reason for its existence and origin.

Summary

◆ **Astrophysics** is the application of physics to **astronomy**. In astrophysics, the laws and principles of physics are used to explain the physical processes that occur inside planets, stars, galaxies and other astronomical objects.

◆ **Gustav Kirchhoff** is acknowledged by many astrophysicists as the 'founding father' of astrophysics. His investigations into the science of spectroscopy prepared the way for many new discoveries with regard to the physical nature of astronomical objects.

◆ The universe in which we live contains a wide variety of astronomical objects of various shapes and sizes. Our knowledge of it has been dramatically increased in recent years with the use of orbiting observatories such as the **Hubble Space Telescope**.

◆ Planet Earth forms part of a system of planets orbiting a star called the Sun. Our Solar System is probably one of many that exist in space. As well as **planets**, the Solar System contains **asteroids**, **comets** and a stream of radiation from the Sun called the **solar wind**.

◆ **Stars** are the main ingredients of the universe. There are many different types of star and many are bound together in **binary systems**. The space between stars is called the **interstellar medium** and contains gas and dust from which stars are made. The interstellar medium forms a **stellar ecosystem** in which stars are born, live and die.

◆ **Galaxies** are collections of stars which are bound together gravitationally. They are grouped together in **clusters**. Our own galaxy the Milky Way forms part of a cluster containing at least 20 galaxies, called the *Local Group*. Clusters are found to form groups called **superclusters**.

◆ The space between galaxies and clusters of galaxies is largely empty. Great **voids** of empty space can be seen and at the largest scales the universe appears to have a **filamental structure**.

◆ Some of the most distant objects are **quasars**. These objects are poorly understood but are thought to be an early kind of galaxy.

◆ The science of astrophysics ranges from the study of comparatively small objects such as planets right up to the large-scale structure of the universe including the detection of radiation from the most distant galaxies. In recent decades, **space astronomy** has extended the range of electromagnetic wavelengths visible to astronomers and many new astrophysical objects have been discovered. This has considerably increased our knowledge and is helping to explain how the origin and structure of the universe came about and what may happen to it in the future.

Questions

1 With whom did astrology originate and how is it different from astronomy? What is the difference between astronomy and astrophysics?

2 **a** Why do astronomers think that the orbit of the Earth has remained stable for as long as life has existed on it?
   b The mass of the Earth is 6.0×10^{24} kg and its radius is 6400 km. Calculate its average density, assuming it is perfectly spherical. If the density of most rocks is between 2000 and 8000 kg m^{-3}, what does your answer suggest that the bulk of the Earth is composed of?

3 **a** What is meant by Terrestrial and Jovian planets? Give two examples of each.
   b Compare the average density of Jupiter (mass $= 1.9 \times 10^{27}$ kg; radius $= 71\ 400$ km) with that of the Earth. (Assume both planets are perfectly spherical.)

4 Outline the main features of the Solar System. In your account you should include:
   a A description of the Sun and its main observational features.
   b The planets, their moons and other bodies that move in the Solar System, illustrating their orbits with a diagram.
   c Ideas about the formation of the Solar System.

5 **a** Outline the basic structure of a star. What are the processes by which energy is transported from the interior to the surface? What is the visible disk of the Sun that we can see from the Earth called?
   b The average distance of the Earth from the Sun is 1.5×10^{11} m. Suppose the Sun suddenly stopped shining. If light travels at 3×10^{8} m s^{-1}, how long would it be before we became aware of this on Earth?

6 What is a variable star? Explain the difference between *intrinsic* variables and *eclipsing* variables. What is a *light curve*?

7 **a** What are *galaxies*? How are they distributed in space?
   b Why does our own galaxy appear to us a faint narrow band of light across the night sky?

8 **a** What is *space astronomy*? How has it extended our knowledge of the universe?
   b In 1972 the Pioneer 10 space probe was launched on a mission to observe Jupiter and Saturn. Moving at a speed of 30 km s^{-1}, it became the first artificial object to escape from the Solar System and head out into interstellar space. Assuming that it was heading for *Proxima Centauri*, a star in the *Alpha Centauri* system situated 4.2 light years from Earth, calculate how many years it would take to get there, travelling at this speed.

9 What are *superclusters*? What appearance does the universe have at the largest scales of distance?

2 Radiation, matter and gravitation

Most of the information we receive from space comes to us in the form of electromagnetic radiation. By analysing this radiation we can understand a great deal about the physical processes causing it. In order to do this we must first understand some basic physics about radiation, matter and motion. In this chapter we will look at some key concepts in physics and dynamics which are important in astrophysics and enable us to learn about the physical properties of the universe.

The nature of electromagnetic radiation

Radiation from all parts of the electromagnetic spectrum reaches the Earth from space, and its detection and analysis is our principal bridge to understanding the processes that govern the universe. The mechanisms that produce this radiation are to do with the nature of matter on an atomic and molecular level. To start with, let's look at some of the properties of electromagnetic radiation and how these may be used to infer certain astrophysical information about the sources producing it.

Electromagnetic radiation may be modelled either in terms of a **transverse wave motion** or as discrete packets of energy called **photons**, and this complementary description (or **wave–particle duality** as it is sometimes called) has been very successful in explaining the varied characteristics of electromagnetic phenomena.

Some properties of electromagnetic radiation such as refraction, diffraction, polarisation and interference effects are best explained using a wave description, while others such as the photoelectric effect, atomic spectra and thermal radiation are better explained using a particle model. In general, if the detection of radiation involves a very large number of photons then a wave description is appropriate. If it is possible to detect individual photons, then a particle explanation is more useful.

In a vacuum, all electromagnetic radiation travels at the speed of light c, equal to 3×10^8 m s^{-1}. The value of c plays an important role in many areas of astronomy and astrophysics, including measures of distance such as the light year, and is also important (as we will see in Chapter 10) in Einstein's Theory of Special Relativity.

The electromagnetic spectrum

The electromagnetic spectrum is shown in Figure 2.1. As we stated earlier, radiating matter can emit light in discrete amounts of energy called photons. A single photon of electromagnetic radiation with frequency f will have an energy given by the **Planck relation:**

$$E_{photon} = hf$$

where E_{photon} is in joules, f is in hertz and h is a fundamental constant called **Planck's constant** which has a value of 6.63×10^{-34} J s.

The relationship between frequency, wavelength λ and speed v for a wave is given by $v = f\lambda$. So by substituting for f in the Planck relation and equating $v = c$, the speed of all electromagnetic waves, we can express the photon energy in terms of its wavelength:

$$E_{photon} = \frac{hc}{\lambda}$$

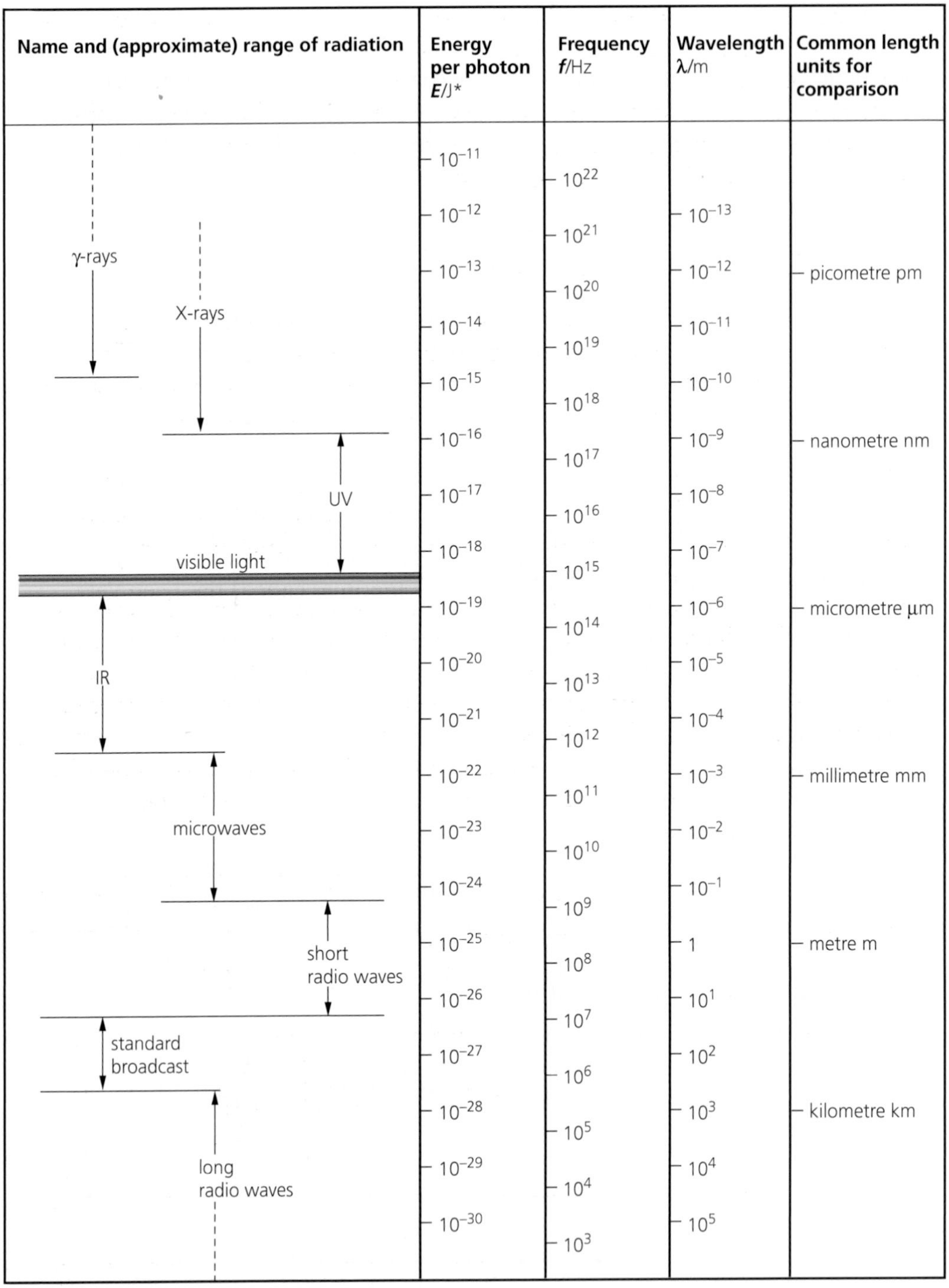

Name and (approximate) range of radiation	Energy per photon E/J*	Frequency f/Hz	Wavelength λ/m	Common length units for comparison
γ-rays	10^{-11}	10^{22}		
	10^{-12}	10^{21}	10^{-13}	
	10^{-13}	10^{20}	10^{-12}	picometre pm
X-rays	10^{-14}	10^{19}	10^{-11}	
	10^{-15}	10^{18}	10^{-10}	
	10^{-16}	10^{17}	10^{-9}	nanometre nm
UV	10^{-17}	10^{16}	10^{-8}	
	10^{-18}	10^{15}	10^{-7}	
visible light				
	10^{-19}	10^{14}	10^{-6}	micrometre μm
	10^{-20}	10^{13}	10^{-5}	
IR	10^{-21}	10^{12}	10^{-4}	
	10^{-22}	10^{11}	10^{-3}	millimetre mm
microwaves	10^{-23}	10^{10}	10^{-2}	
	10^{-24}	10^{9}	10^{-1}	
	10^{-25}	10^{8}	1	metre m
short radio waves	10^{-26}	10^{7}	10^{1}	
standard broadcast	10^{-27}	10^{6}	10^{2}	
	10^{-28}	10^{5}	10^{3}	kilometre km
long radio waves	10^{-29}	10^{4}	10^{4}	
	10^{-30}	10^{3}	10^{5}	

*E can be expressed relative to the electronvolt by dividing these numbers by 1.6×10^{-19}

Figure 2.1 The electromagnetic spectrum

Notice that the photon energy is directly proportional to the frequency but inversely proportional to the wavelength of the radiation. This explains the form of the electromagnetic spectrum in Figure 2.1. For high energies we have high frequencies but short wavelengths; for low energies, low frequencies and long wavelengths.

The energy of a single photon of visible light is very small. For red light of wavelength 630 nm (1 nm = 10^{-9} m) we have

$$E_{photon} = \frac{(6.63 \times 10^{-34}\,\text{J s}) \times (3.0 \times 10^{8}\,\text{m s}^{-1})}{(630 \times 10^{-9}\,\text{m})} = 3.1 \times 10^{-19}\,\text{J}$$

Since this is such a small number, it is useful to define another unit called the **electronvolt** (eV) as a more convenient measure (see Box 2.1, page 34). The electronvolt is defined as the energy acquired by an electron in moving through a potential difference of one volt. As energy = charge × voltage change, $1\,\text{eV} = 1.602 \times 10^{-19}\,\text{C} \times 1\,\text{V} = 1.602 \times 10^{-19}\,\text{J}$.

Photons can have a wide range of energies. Gamma rays have energies of the order of mega-electronvolts (MeV), whereas X-rays have energies of kilo-electronvolts (keV). Radio photons have very small energies, typically about 10^{-10} eV.

The atmosphere of the Earth is only transparent to electromagnetic radiation over certain ranges of wavelengths. There is an 'optical window' and a 'radio window' through which radiation from these parts of the spectrum can reach the Earth's surface, together with a certain amount of infrared (Figure 2.2). All other radiations are effectively blocked by the Earth's atmosphere which is fortunate for us as many of them are harmful to life. This is however a hindrance to astrophysicists who want to study the whole of the electromagnetic spectrum and is why space astronomy has revolutionised our understanding of the universe by enabling the full range of wavelengths to be explored.

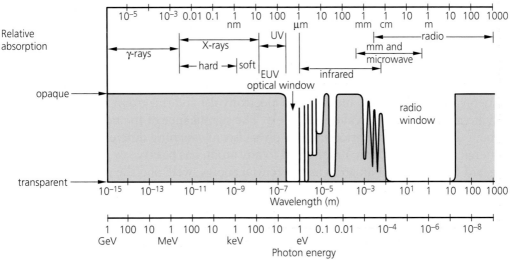

Figure 2.2 Relative absorption of electromagnetic radiation by the Earth's atmosphere. The optical and radio 'windows' can be clearly seen

Atomic structure

An atom is composed of protons, neutrons and electrons and has an average size of about 10^{-10} m. Every atom consists of a positively charged **nucleus** of approximate diameter 10^{-14} m, containing **nucleons** made up of a number of protons and neutrons. The number of protons is called the **atomic** or *proton number* and given the letter Z while the total number of nucleons in the nucleus is called the **mass** or **nucleon number** and given the letter A.

Surrounding the nucleus are negatively charged electrons which are arranged in electron orbits or **electron shells** of definite energies. Usually, the number of positive charges in the nucleus and the number of negative charges in the electron shells are equal and, as a consequence, the atom is electrically neutral and has no net charge.

The Bohr model

The **Bohr model** of the hydrogen atom provides us with a useful picture that helps us to understand how electromagnetic radiation can be emitted and absorbed by an atom. In this model, put forward by the Danish physicist Neils Bohr (1885–1962), the electrons move around the nucleus in well defined orbits that correspond to specific energy levels. Bohr used ordinary mechanics to describe the orbit of the electron around the hydrogen nucleus but he added a number of important conditions or **postulates**.

Postulate 1. The energy of the electron in its orbit is *constant*.

If the energy were not constant then the electron would radiate energy away and eventually spiral in to the nucleus. The electron in its orbit has both kinetic energy by virtue of its orbital motion and potential energy since it is attracted to the nucleus by electrostatic forces.

Postulate 2. The electron can only occupy certain definite orbits and cannot exist 'in between' them.

Since the electron in its orbit has a certain energy, we associate with it a specific **energy level** corresponding to the particular orbit it happens to be in at the time. The lowest energy orbit that an electron has in an atom is called its **ground state**. Energy levels are usually measured in electronvolts and the ground state for the hydrogen atom has the value -13.6 eV. The significance of the negative sign comes about due to the convention used by physicists in assigning differences in electron energies. A free electron, unattached to any atom, has positive or *greater* energy. We regard the interior of an atom as a potential energy 'well' and the potential energy at its 'surface' to be zero. Inside the atom at some distance from the nucleus the electron therefore has negative or *less* potential energy. This is purely a convention but beware of confusion here! The further away from the nucleus the electron is, the *greater* is its electrical potential energy.

Postulate 3. The absorption or emission of electromagnetic radiation in an atom comes about due to a *transition* of an electron between orbits.

When a photon is **absorbed** by an atom, the electron makes a transition from a lower energy orbit to a higher one. When an atom **emits** a photon the electron makes a transition from a higher to a lower energy orbit. It is important to understand that when atoms emit or absorb light, then the principle of energy conservation implies that they can *only* do so if the photon's energy is equal to the *difference* in two specific energy levels. The energy of the emitted/absorbed photon is said to be **quantised** and is equal to the difference in energy between the two orbits. If an electron makes a transition from an orbit of higher energy E_2 to an orbit of lower energy E_1 then the emitted photon will have a frequency given by the Planck relation:

$$E_2 - E_1 = hf \qquad \text{or} \qquad f = \frac{E_2 - E_1}{h}$$

When the electron is in the ground state it is in its lowest possible energy level, and the atom is then in its lowest energy state. If energy is added to the atom either by heating it or by the absorption of photons then an electron can jump to higher energy orbits and the atom is said to be in an **excited state**. If the electron gains enough energy, then it can escape from the nucleus altogether and the atom is said to be **ionised**.

Notice that in order to remove the electron completely from the hydrogen atom in its ground state, an energy of at least 13.6 eV is required. This is called the **ionisation energy** for hydrogen. Once the electron has escaped it can have a continuous range of energies. Only while it is inside the atom is its energy quantised.

The Bohr model was able to explain many of the features observed in the spectrum of atomic hydrogen. The hydrogen spectrum has a sequence of spectral lines that start at 656.31 nm and ends at 364.56 nm. These lines at visible wavelengths are called the **Balmer lines**, and correspond to electrons dropping down to the lowest excited state from higher energy levels. The entire sequence is called

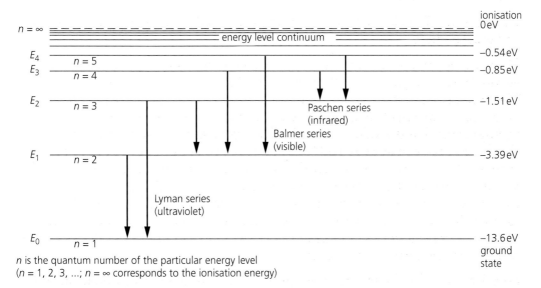

n is the quantum number of the particular energy level
(n = 1, 2, 3, ...; $n = \infty$ corresponds to the ionisation energy)

Figure 2.3 Energy levels and spectral lines for hydrogen according to the Bohr model

the **Balmer series**. Lines produced by electron transitions to the ground state are all emitted at ultraviolet wavelengths and are called the **Lyman series**, while those to the second excited state are emitted in the infrared and are called the **Paschen series**. The energy levels for hydrogen showing the variety of electron transitions and resulting spectral lines are shown in Figure 2.3.

It must be stressed that the Bohr model does not provide a complete picture of atomic spectra. While the model can explain the main features of the spectrum of hydrogen and its ionisation energy, it cannot account for the spectra of atoms with more than one electron. It did however point to the fact that electrons exist in definite energy levels within atoms and that electron transitions give rise to spectral lines, and the Balmer series turns out to be of great importance in classifying stars by their spectral characteristics.

The quantum mechanics model

A more complete description of atomic structure is provided by the modern theory of **quantum mechanics**. Quantum mechanics retains the idea of discrete electron energy levels, but does away with the idea that motion of the electrons can be described using Newtonian mechanics. It is more correct to talk of collective **quantum states** and energy levels but not orbits. Quantum mechanics says that we cannot know the exact position of an electron in an atom at any specific time. The electrons are to be found in a diffuse **electron cloud** that surrounds the nucleus, with the Bohr orbit representing the location of the highest probability of the position of the electron. When a photon is emitted or absorbed then the transition is between quantum states with each state having an assigned energy level.

Quantum mechanics has been very successful in explaining the spectra of atoms with more than one electron, but the theory is not so easy to visualise and the mathematics involved is beyond the scope of this book.

Other sources of electromagnetic radiation

Electromagnetic radiation is also produced when electric charges are accelerated by electric fields, including when they are oscillating. This effect was predicted by James Clerk Maxwell (1831–79) and we are commonly familiar with it as the means of propagation of radio waves from radio transmitters. Electrons spiralling in magnetic fields also produce electromagnetic radiation (this is how a microwave oven works). The hot ionised gases associated with many astronomical objects moving under the effects of gravity also radiate in these ways.

Finally, another effect is that of **fluorescence** where high energy photons are converted into ones of lower energies. An example is when an ultraviolet photon is absorbed by an atom, exciting an electron to a higher energy level. The electron then makes a sequence of transitions back to the ground state, emitting lower frequency light at each step of the process. Fluorescence is commonly employed in the 'ghost train' at amusement parks to produce eerie effects of blue and violet light in the shapes of apparitions. Fluorescence is an important process in **emission nebulae** (Chapter 6).

Thermal radiation – Stefan's Law

Suppose we take a metal bar and heat it with a blowtorch. At first the bar will glow a dull red. As it becomes hotter, the bar will change colour from red through to orange and then yellow and, if it gets extremely hot (and could be prevented from melting), to a brilliant bluish-white. This suggests that as an object reaches higher temperatures, it emits radiation of shorter wavelengths.

Why does the bar change colour when it increases in temperature? The Austrian physicist Josef Stefan (1835–93) carried out a series of experiments which showed that a body emits energy at a rate that is related to its temperature. This empirical relationship was also derived theoretically by another physicist, Ludwig Boltzmann (1844–1906), using thermodynamic assumptions about atoms and molecules. They were led to conclude that *a body when heated, will emit electromagnetic radiation over a range of wavelengths with a total intensity that is proportional to the fourth power of its temperature.* In addition, *the wavelength of the peak emission intensity is inversely proportional to the temperature of the body.*

The energy radiated by a body can be usefully defined in terms of the **energy flux**, i.e. the amount of energy leaving unit area in unit time. We therefore obtain the **Stefan–Boltzmann** or **Stefan's Law** as it is more commonly known:

$$P = \sigma T^4$$

where P is the radiated power per unit area in J m^{-2} s^{-1} and T is the temperature in kelvin. The constant of proportionality σ is called the **Stefan–Boltzmann constant** and is equal to 5.67×10^{-8} W m^{-2} K^{-4}. The total power radiated by a body of area A and temperature T is therefore $A\sigma T^4$.

Stefan's Law holds true for a body that is in **thermal equilibrium**. This means that, for the temperature to remain steady, the rate of energy absorbed by the body from its surroundings must equal the rate of energy flowing out from it, and the temperature at which this balance is maintained is called the **equilibrium temperature**. This is expressed more formally as **Kirchhoff's Law**: *for any given temperature, the ratio of the capacity of a body to emit radiation to its capacity to absorb it (at a particular wavelength) is constant and is independent of the composition of the body.*

This statement implies that *a good emitter of radiation must also be a good absorber.* If a body is an efficient absorber of radiation at a given wavelength, then it will also be an efficient radiator at that wavelength. This principle is very important in stellar spectroscopy since it is found that atoms and molecules in the hot atmosphere of a star can be identified by their line spectra and, by Kirchhoff's Law, the gas that *emits* bright spectral lines should at the same temperature *absorb* energy at the same wavelength. We will be looking at the spectra of stars and Kirchhoff's rules of spectral analysis in Chapter 4.

The blackbody – Wien's Law

It was shown by Boltzmann that the Stefan–Boltzmann Law is only valid for a body that is a perfect absorber of energy. Such an object would therefore appear absolutely black at all wavelengths and is known as a **blackbody**. You can make a good approximation to a blackbody by taking a hollow sphere and drilling a small hole in the side of it. The hole then behaves like a blackbody. Any radiation entering the hole from outside will become trapped, undergoing repeated reflections inside the sphere, until eventually it is absorbed by the inner surface.

A blackbody does not reflect any light, which means that the energy flux it emits depends *only* on its temperature and not on its composition, in accordance with Stefan's Law.

The Sun and other stars emit radiation very much like an ideal blackbody would. Intuitively this seems rather strange. Why are they called 'black' when they most obviously are not. We have to understand that stars are blackbodies because they absorb light at any wavelength but do not reflect any back (if you were to shine a beam of light at the Sun, it would not be reflected back to you).

A blackbody emits electromagnetic radiation over a wide range of wavelengths but there will be one wavelength called the **peak wavelength** where the emission of radiation has its *maximum* intensity. The German physicist Wilhelm Wien (1864–1928) discovered a simple relationship between the absolute temperature T of a blackbody and the peak wavelength λ_{max} (in metres) at which the radiated energy reaches its maximum intensity:

$$\lambda_{max} = \frac{2.90 \times 10^{-3}}{T}$$

This relationship is known as **Wien's Law**. Since the wavelength is inversely proportional to the temperature, the dominant wavelength of a blackbody radiator *must decrease* as it gets hotter, just as we observe when we heat the metal bar. An object at room temperature (for example 300 K) emits mainly infrared radiation. A very cold object of temperature a few kelvin above absolute zero would emit primarily microwaves whereas a body of a few million kelvin would emit at X-ray wavelengths (see Table 2.1, Box 2.1, page 34).

When a blackbody, initially in equilibrium at a certain temperature, is provided with extra energy to absorb, it heats up until it reaches a new equilibrium temperature at which it can radiate away all the incoming radiation at the same rate as it is absorbed. It will emit at all wavelengths even though the peak wavelength may not be visible to the human eye. In Figure 2.4, the pink band represents the total energy radiated per m² per second for a blackbody of temperature 3000 K in a band of wavelengths between λ and $\lambda + \Delta\lambda$.

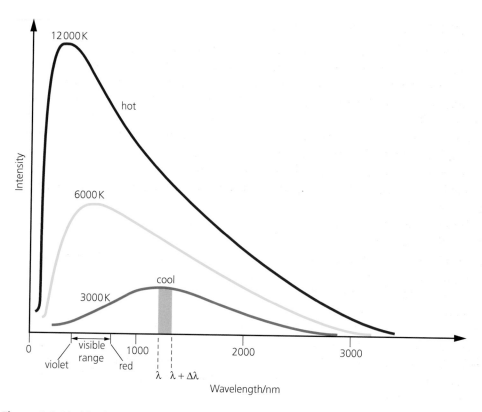

Figure 2.4 Blackbody curves for objects at different equilibrium temperatures

The intensity distribution of blackbody radiation always has a characteristic shape – a graph showing the energy flux with wavelength is a *continuous* one and does not contain any characteristic lines. Figure 2.4 shows the spectra for blackbody radiators at different temperatures. Notice that the higher the temperature, the shorter the wavelength of maximum intensity, just as we would expect from Wien's Law.

These curves are known as **Planck** or **blackbody curves**. The important point to realise when you see a blackbody curve is that the processes that give rise to the emission of radiation depend *only* on temperature and not on any other property such as the chemical composition of the object. The total area under a blackbody curve is equal to the total energy emitted per square metre per second over all wavelengths.

Measurements made above the Earth's atmosphere of the intensity distribution of sunlight over a broad range of wavelengths show that the Sun is a good approximation to a blackbody when compared with the theoretical blackbody curve at a temperature of 5800 K, and it is because stars are so much like blackbodies that astrophysicists are able to deduce their surface temperatures in this way.

WORKED EXAMPLE 2.1

The average flux of energy reaching the Earth from the Sun is called the **solar constant** and its value is 1370 W m^{-2}. The Sun emits electromagnetic radiation over a wide range of wavelengths but its intensity is found to peak at about 500 nm. Assuming the Sun radiates like a blackbody, use **a** Stefan's Law and **b** Wien's Law to calculate its surface temperature. The Stefan–Boltzmann constant $\sigma = 5.7 \times 10^{-8}$ W m^{-2} K^{-4}. The radius of the Sun can be taken as 7×10^8 m and the distance of the Earth from the Sun as 1.5×10^{11} m.

a If T is the temperature of the Sun's surface and r_S its radius, then using Stefan's Law (and remembering that the total surface area of a sphere is $4\pi r^2$) the total power radiated will be:

$$P_{tot} = A\sigma T^4 = 4\pi r_S^2 \times \sigma \times T^4 \text{ W}$$

At the distance of the Earth from the Sun r_{ES}, this power is spread over a sphere of radius r_{ES} (Figure 2.5). Since the area of this sphere is $4\pi r_{ES}^2$, the energy falling per second on unit area of the Earth (the solar constant) is given by:

$$\frac{1}{4\pi r_{ES}^2} \times 4\pi r_S^2 \sigma T^4 = \frac{r_S^2}{r_{ES}^2} \sigma T^4 = 1370 \text{ W m}^{-2}$$

We are given $r_S = 7 \times 10^8$ m, $r_{ES} = 1.5 \times 10^{11}$ m and $\sigma = 5.7 \times 10^{-8}$ W m^{-2} K^{-4}, so:

$$T^4 = \frac{1370 \text{ W m}^{-2}}{(5.7 \times 10^{-8} \text{ W m}^{-2} \text{ K}^{-4})} \times \frac{(1.5 \times 10^{11} \text{ m})^2}{(7 \times 10^8 \text{ m})^2}$$

from which we can calculate the surface temperature of the Sun to be:

$$T \approx 5800 \text{ K}$$

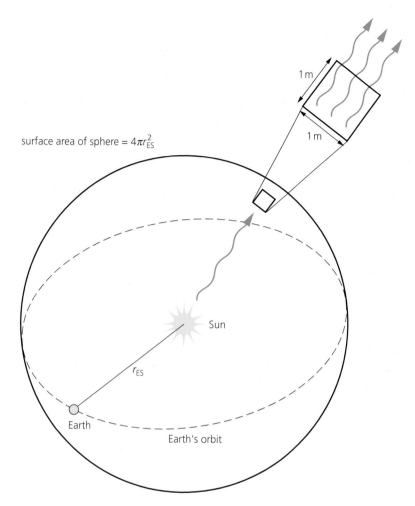

surface area of sphere = $4\pi r_{ES}^2$

1 m

1 m

Sun

r_{ES}

Earth

Earth's orbit

energy falling per second per m² on sphere of radius r_{ES}
= energy falling per second on unit area of the Earth
= the solar constant 1370 Wm⁻²

Figure 2.5

b   The Sun's radiation peaks at about 500 nm = 500×10^{-9} m. Using Wien's Law, the surface temperature of the Sun is:

$$T = \frac{2.90 \times 10^{-3}}{500 \times 10^{-9}} = 5800 \text{ K}$$

(Remember that this is the temperature of the Sun's surface or **photosphere**. The temperature inside the Sun and in its atmosphere are very different and cannot be deduced by these equations.)

Box 2.1 The electronvolt – a unit of photon energy

We have seen that the energies of photons are given by the Planck relation $E_{photon} = hf = hc/\lambda$. For example, a photon of visible red light with $\lambda = 630$ nm has an energy of:

$$\frac{(6.67 \times 10^{-34} \text{ J s}) \times (3.0 \times 10^8 \text{ m s}^{-1})}{(630 \times 10^{-9} \text{ m})} = 3.2 \times 10^{-19} \text{ J}$$

Similarly, BBC *Radio 4* broadcasts radio photons of wavelength 3.3 m and energy 6×10^{-26} J, whereas in the gamma ray region of the electromagnetic spectrum a photon of frequency 10^{20} Hz would have an energy of $(6.6 \times 10^{-34} \text{ J s}) \times (10^{20} \text{ Hz}) = 6.6 \times 10^{-14}$ J.

What you will notice is that if we measure the energies of photons in **joules** then we get very small quantities. In order to make the numbers more manageable, physicists use a more convenient unit of energy – the **electronvolt**.

One electronvolt (1 eV) is the energy gained by an electron when it is moved through a potential difference (p.d.) of one volt. Since the work done in moving a charge Q through a p.d. of V is given by $Q \times V$,

$$1 \text{ eV} = \text{electron charge} \times 1 \text{ V} = 1.6 \times 10^{-19} \text{ C} \times 1 \text{ V} = 1.6 \times 10^{-19} \text{ J}$$

Converting energy values in joules to electronvolts is easy – just divide by the value of the electron charge. For example, the energy of the photon of red light in electronvolts is:

$$\frac{3.2 \times 10^{-19} \text{ J}}{1.6 \times 10^{-19} \text{ C}} = 2 \text{ eV}$$

Conversely, to convert electronvolts to joules, multiply by the value of the electron charge. For example, the energy in joules of an X-ray photon of 60 keV is:

$$(60 \times 10^3 \text{ eV}) \times (1.6 \times 10^{-19} \text{ C}) = 9.6 \times 10^{-15} \text{ J}$$

The electronvolt is not an SI unit, but it is widely used in atomic and nuclear physics. Astronomers detect and measure photon energies in order to gain information about astrophysical processes. Some common multiples are:

kilo-electronvolts (keV) 1 keV = 1000 eV
mega-electronvolts (MeV) 1 MeV = 1 000 000 eV
giga-electronvolts (GeV) 1 GeV = 1 000 000 000 eV

Table 2.1 shows the range of photon energies in electronvolts for various parts of the electromagnetic spectrum together with their wavelengths and corresponding blackbody temperatures.

Table 2.1

	Photon energy/eV	λ/m	**Blackbody temperature/K**
Radio	$<10^{-5}$	>0.1	<0.03
Microwave	10^{-5}–10^{-2}	0.1–10^{-4}	0.03–30
Infrared	0.01–2	10^{-4}–7×10^{-7}	30–4100
Visible	2–3	7×10^{-7}–4×10^{-7}	4100–7300
Ultraviolet	3–1000	4×10^{-7}–10^{-9}	7300–3×10^6
X-ray	1000–10^5	10^{-9}–10^{-11}	3×10^6–3×10^8
Gamma ray	$>10^5$	$<10^{-11}$	$>3 \times 10^8$

The Doppler effect

You may have noticed when standing at a station that when a train is coming towards you, the note of its klaxon is higher and then drops in frequency as it passes by and starts to recede. This is an example of the **Doppler effect**, named after the Austrian physicist Christian Doppler (1803–53).

The reason why this happens is that when the train is approaching, more sound waves per second are reaching your ears than if the train is stationary and the wavelength becomes shortened as a result. Since for a given wave speed, the frequency is inversely proportional to the wavelength, the frequency of the sound is increased. As the train recedes from you, there are less sound waves per second reaching your ears, the wavelength is effectively longer and so the frequency of the sound is lower.

The same thing happens with electromagnetic radiation. With light, depending on the motion of the object with respect to the observer, the colour of the light is affected. An approaching light source is a little more blue (shorter wavelength) than it would otherwise be, and one that is moving away is a little redder (longer wavelength).

The Doppler effect for light can be used to estimate the speed at which distant stars are receding from the Earth. Suppose a star moving away from the Earth with a velocity v (much lower than the speed of light, c) emits light at a wavelength λ_0 and frequency f_0 Hz. Then in one second f_0 waves are emitted. The light travels outwards at the speed of light c and because the star is receding at a velocity v, the f_0 waves will occupy a distance $1 \text{ s} \times (c + v) \text{ m s}^{-1} = (c + v)$ metres. The observer on Earth therefore sees an **apparent wavelength** λ (since $v = f\lambda$) of:

$$\lambda = \frac{(c + v)}{f_0} = \frac{(c + v)}{c}\lambda_0 = \left(1 + \frac{v}{c}\right)\lambda_0$$

Rearranging, we get:

$$\frac{\lambda - \lambda_0}{\lambda_0} = \frac{v}{c}$$

$$\frac{\Delta\lambda}{\lambda_0} = \frac{v}{c}$$

where $\Delta\lambda = \lambda - \lambda_0$ is the difference between the wavelengths λ and λ_0. This expression is only valid when v is much less than c since it ignores the effects of special relativity. Put another way, $\Delta\lambda$ must be much less than λ_0 for the expression to be valid.

Red shifts and blue shifts

If λ is greater than λ_0 then the star is moving *away* from the Earth and v/c is positive. We say that the light has been **red-shifted** towards the red end of the electromagnetic spectrum. If λ is less than λ_0 then the star is moving *towards* us and light has been shifted towards the blue end or **blue-shifted**. In this case v/c is negative.

By comparing the wavelengths in the spectral lines of a star with those in a laboratory on Earth, astrophysicists can measure the difference in wavelengths $\Delta\lambda$, calculate the velocity and tell whether the star is moving towards or away from us. Notice that the Doppler effect allows measurement only of the velocity *along* the line of sight (radial velocity) and says nothing about the velocity *across* the line of sight (tangential velocity).

WORKED EXAMPLE 2.2

In a laboratory, a certain spectral line in the hydrogen spectrum is observed to have a wavelength of 656.285 nm. An astrophysicist measures the same line in the spectrum of the star *Vega* as 656.255 nm. Is *Vega* moving towards us or away from us and what is its velocity in this direction?

The wavelength shift is $\Delta\lambda = \lambda - \lambda_0 = (656.255 - 656.285)$ nm $= -0.030$ nm. The minus sign indicates that it is a blue shift.

$$\frac{\Delta\lambda}{\lambda_0} = \frac{v}{c} \quad \text{so} \quad v = c\frac{\Delta\lambda}{\lambda_0} = (3 \times 10^8)\frac{(-0.030)}{(656.285)} = -13.7 \text{ km s}^{-1}$$

The star *Vega* is moving *towards* us with a velocity of 13.7 km s^{-1}.

The Doppler formula applies both to the local motions of stars and to the universe as a whole. As will be explained in Chapter 10, the universe is expanding and the red shifts of galaxies and other distant objects are interpreted as due to the expansion of space itself. Astronomers attach the symbol z to the Doppler formula (especially when it is used in this way) so that

$$z \equiv \frac{\Delta\lambda}{\lambda_0} = \frac{v}{c}$$

and a distant object is quoted in terms of its 'z'. The higher the value, the faster the object is receding from us. However, at very large distances, the speed of an object is so large that the Doppler formula must be corrected for the effects of special relativity (see Chapter 10). QSOs, being some of the most distant objects detected, have z values of nearly 5.

Motion under gravity

By studying the motions of stars, galaxies and other astronomical objects, astrophysicists can learn much about their organisation and physical arrangement. Some of these motions are regular and periodic such as the motion of a planet around the Sun or two stars in a binary orbit, whereas others such as the ejection of matter from a supernova are projectile-like and random.

In almost all cases it is **gravity** that plays the key role in the motions of astronomical objects. Although gravity is the weakest force compared with the electrostatic and nuclear forces which control matter at the atomic and nuclear level, it dominates the universe at visible scales and operates over the greatest range of distances. Gravity is responsible for the fact that we feel weight, for the orbital motion of the Moon about the Earth, the Earth about the Sun, and the Sun about the galactic centre, and even for the orbital motion of clusters of galaxies. Later, in Chapter 10, we will see that gravity plays an important part in determining the ultimate fate of the universe.

The study of celestial motions, or **celestial mechanics** as it is still called, was begun by Sir Isaac Newton (1642–1727) with his Laws of Motion and Universal Gravitation, although important analysis of the motions of the planets in the Solar System had been made earlier by Johannes Kepler (1571–1630), based on observations by the Danish astronomer Tycho Brahe (1546–1601). Celestial mechanics is also important in its modern application to communication and navigation satellites, interplanetary probes and the Apollo Moon landings, where the use of Newton's and Kepler's laws is essential in determining the orbital dynamics of interplanetary space flight. This application of celestial mechanics to both natural and artificial objects has led to the term **astrodynamics** being introduced, which covers the motion of all objects moving in space under the influence of gravity.

Astrodynamics and the two-body problem

Astrodynamics can describe the motions of many bodies moving in gravitational fields; however, for our purposes we will consider only the mechanics of *two* bodies that interact gravitationally. In astrodynamics this is commonly known as the **two-body problem**, which poses the question: *given at any instant the positions and velocities of two masses moving under their mutual gravitational attraction, can we calculate their positions and velocities at any other time past or future?*

In the two-body problem there are three sets of physical laws that are used to describe the motion of the two bodies moving under gravity:

1 Kepler's Laws
2 Newton's Laws of Motion
3 Newton's Law of Gravitation

Kepler's Laws

By analysing the observations of the planets made by Tycho Brahe, Kepler found a relationship between the sizes of the planets' orbits and their periods of revolution around the Sun. His most important finding was that the planets move in **ellipses** with the Sun at one of the **foci**. An ellipse is a curve such that at each point on the curve the sum of the distances to two given points (the foci) is constant. The line through the foci to the ends of the ellipse is called the **major axis**, $2a$, and the line perpendicular to this the **minor axis**, $2b$ (Figure 2.6). In the Solar System, the planets (with the exception of Pluto) move in ellipses which are very nearly circular ($a \approx b$) but comets move through the Solar System in highly elliptical orbits ($a > b$).

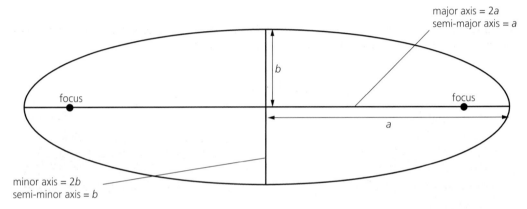

Figure 2.6 Basic geometry of an ellipse. A circle is a special case of an ellipse when the two foci coincide

Kepler also found that a line drawn from a planet to the Sun sweeps out equal areas in equal times. This means that the velocity of a planet around the Sun is not uniform but varies according to its distance. The further away it is the slower it moves and as it approaches the Sun it speeds up. This is a consequence of gravity. As the planet approaches the Sun the gravitational attraction pulls the planet towards it and then slows it down as it moves away.

Finally, Kepler deduced a mathematical relationship between the period of a planet and its **semi-major axis** (one half of the major axis).

These findings together are known as **Kepler's Laws of Planetary Motion** and are summarised below:

1st Law *The orbit of a planet is an ellipse with the Sun situated at one of the foci while the other focus is empty* (Figure 2.7a).

2nd Law This is known as the **Law of Equal Areas**. *A straight line (called the radius vector) connecting the Sun and a planet will sweep out equal areas in equal times* (Figure 2.7b).

3rd Law *The squares of the orbital periods of the planets are proportional to the cubes of their semi-major axes.* If a planet has a period T and semi-major axis a (or a radius r if the orbit is circular) then this may be expressed mathematically as:

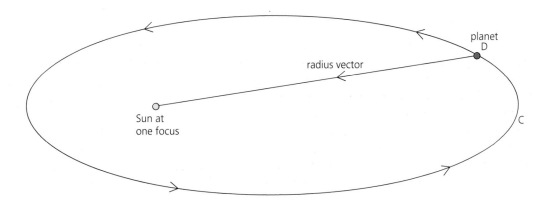

Figure 2.7 (a) Planets in the Solar System move along the path of an ellipse with the Sun at one focus. The line connecting the Sun and a planet is called the radius vector

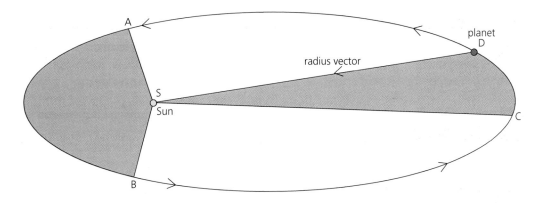

Figure 2.7 (b) The Law of Equal Areas: area ASB = area CSD. The planet moves from A to B in the same time it takes to move from C to D

$$\frac{T^2}{a^3} = \text{constant}$$

The value of the constant depends on the mass of the Sun.

The second and third laws imply that the force acting between the Sun and a planet is a central force and decreases with distance. However, Kepler was unable to explain what the nature of this force was and it remained for Newton to supply the answer to exactly what kept the planets in orbit around the Sun.

Although Kepler's Laws were originally applied to the motion of the planets, they are also obeyed by stars orbiting in a binary system, stars orbiting around a galaxy and even galaxies in orbit around each other! Kepler's Laws are therefore of great importance in determining the periods and separations of stellar systems.

Newton's Laws of Motion

Newton's Laws of Motion are the foundation of dynamics. You will certainly have met them before and they are summarised below:

1st Law   *A body continues in its state of rest or uniform motion unless acted upon by an external force.*

In the case of planets orbiting the Sun there must be a force causing them to move in an ellipse or otherwise they would just move off at a tangent in the same direction forever! Newton is saying that a continuously uniform motion is the natural state of an object in the universe and any deviation from this state implies that a force is acting somewhere, even if it isn't immediately obvious where the force originates.

2nd Law   *Force is equal to the rate of change of momentum or, alternatively, the rate of change of velocity (acceleration) of a body is proportional to the force acting on it.*

What this is saying is that if a body is observed to accelerate there must be a net force acting on it. The constant of proportionality is what we define as inertial mass so that mathematically we can write the law as *force = mass × acceleration*. From Kepler's Laws the planets speed up as they approach the Sun and slow down as they move away, so, by Newton's 2nd Law, there must be a force acting on them.

3rd Law   *If a body A exerts a force on body B then body B exerts an equal but opposite force on body A.*

This third law implies that forces act in pairs and are equal and opposite. For example the Moon pulls on the Earth with the same force as the Earth pulls on the Moon but in the opposite direction. However, because of their difference in mass, their accelerations towards each other are different.

Newton's Law of Gravitation

Whether or not an apple fell on his head, the genius of Newton was to realise that *every body in the universe attracts every other body by means of the gravitational force.*
  By applying his Laws of Motion to Kepler's Laws, Newton showed that it was gravity which was responsible for keeping a planet in its orbit and that it was a centrally directed force along the line connecting the Sun and the planet. Furthermore, he showed that if the focus of the ellipse is the centre of motion, then the force must take the form of an inverse square law and from his second and third laws he deduced the **Universal Law of Gravitation**:

Every body in the universe attracts every other body with a force the magnitude of which is proportional to the product of the masses of the bodies and inversely proportional to the square of the distance between them. The direction of the force is along the line connecting the two bodies.

Mathematically, if we have two masses M_1 and M_2 separated by a distance r, then the gravitational force F acting between them is

$$F \propto \frac{M_1 M_2}{r^2} \qquad \text{or} \qquad F = G \frac{M_1 M_2}{r^2}$$

where G is the **universal constant of gravitation** equal to 6.67×10^{-11} N m^2 kg^{-2}.

Gravity only attracts and never repels. The force depends on the *mass* of the objects, not on what they are made of. Another consequence of the Law of Gravitation is that it is able to explain the observation made by Galileo that all bodies close to the Earth's surface fall at the same rate. By equating Newton's 2nd Law with the Law of Gravitation for a mass m falling towards the Earth of mass M_E (where r_E is the Earth's radius) we have:

$$F = ma = \frac{GM_E m}{r_E^2}$$

or

$$a = \frac{GM_E}{r_E^2}$$

This value of a is of course g, the acceleration due to gravity, equal to 9.81 m s^{-2}. Notice that g does not depend on the mass of the falling object but only on the mass of the Earth and its radius.

Mass and weight

Don't get mass and weight confused! When you stand on a scale to measure your weight, the scale reads the attractive force between you and the Earth. This force, your *weight*, should be recorded in Newtons but the scale is calibrated in kg. Your *mass* is related directly to the number and type of particles that make up your body. It is the total amount of matter which makes up the real you. This is *not* a force.

WORKED EXAMPLE 2.3

The acceleration due to gravity on the Moon is approximately $\frac{1}{6}$ of the value on the Earth. How much would a woman of mass 70 kg weigh on the Moon?

Using Newton's 2nd Law, the woman's weight would be 70 kg $\times \frac{1}{6} \times$ (9.81 m s^{-2}) = 1.1×10^2 N. This compares with her weight on Earth of 6.9×10^2 N.

So if you want to lose weight go to a planet with a lower acceleration due to gravity – not that you would look any different, as your mass would remain the same!

Gravity and orbits

The circular orbit

A special case of orbital motion is the circular orbit (Figure 2.8). We can use ideas from circular motion (see Box 2.2) to derive an expression for the radius of the orbit of a planet of mass M_p about the Sun of mass M_S in terms of its orbital period, assuming that the orbit is a circular one with a radius r.

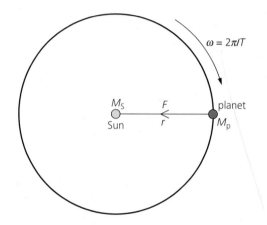

Figure 2.8 The circular orbit. The centripetal force is provided by gravity

At each point of the orbit, the force between the Sun and the planet F (gravity) acts as a centripetal force so we can write:

$$F \text{ (gravity)} = F \text{ (centripetal)}$$

or

$$\frac{GM_SM_p}{r^2} = M_p r \omega^2$$

where ω is the angular velocity in rad s^{-1} of the planet around the Sun. Cancelling the M_p on both sides and rearranging we get:

$$\omega = \sqrt{\frac{GM_S}{r^3}}$$

and since $\omega = 2\pi/T$ where T is the period we obtain:

$$\frac{T^2}{r^3} = \frac{4\pi^2}{GM_S}$$

As the product GM_S is constant for any planet, T^2/r^3 is also constant which is exactly what we would expect from Kepler's 3rd Law. We have derived the third law for the circular orbit; more advanced mathematics gives the same result for an elliptical orbit.

Box 2.2 Circular motion

A body moving in a circle with **uniform speed** will experience a force acting towards the centre called the **centripetal force** (the word centripetal means 'towards the centre').

Figure 2.9a shows a body moving in a circle of radius r. The body travels equal distances along the perimeter of the circle in equal times but the *direction of its velocity is changing*. At A and B the speed is constant but the *velocity* of the body has changed direction (remember that velocity is a *vector* quantity). So although the body is moving with constant speed, its velocity is continually changing. In physics we define a change in velocity as an **acceleration** and by Newton's 2nd Law there *must* be a force acting on the body.

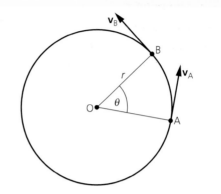

Figure 2.9 (a) A body in uniform circular motion has constant speed but changing velocity

We define the **angular velocity** ω as the change in angle per second, θ/T where θ is in radians, so that $\theta = \omega t$. If s is the length of the arc AB, then using the formula for the arc length of a circle, $s = r\theta$, we can write $s/t = r\theta/t$. Now the *speed* of the body v is s/t and so $v = r\omega$.

Figure 2.9a shows the body at A and B with corresponding velocity vectors $\mathbf{v}_A$ and $\mathbf{v}_B$ which are tangents to the circle at A and B. If we consider a small time interval Δt in the body's rotation about O then the points A and B will be separated by a small corresponding angular displacement $\Delta\theta$. We can now draw a vector triangle XYZ that represents the *change in velocity* $\mathbf{v}_B - \mathbf{v}_A$ (Figure 2.9b). In XYZ, the direction of YZ represents the velocity change $\mathbf{v}_B - \mathbf{v}_A$ and the magnitude of $\mathbf{v}_A$ and $\mathbf{v}_B$ is v. Now when $\Delta\theta$ is small we can use the arc length formula to write the magnitude of $\mathbf{v}_B - \mathbf{v}_A$ as:

$$| \mathbf{v}_B - \mathbf{v}_A | = v\Delta\theta$$

and since acceleration a is change in velocity in unit time, we have:

$$a = \frac{v\Delta\theta}{\Delta t} = v\omega$$

Notice also that since $v = r\omega$, we can write:

$$a = \frac{v^2}{r} \qquad \text{or} \qquad a = r\omega^2$$

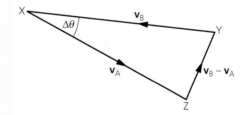

Figure 2.9 (b) The velocity vector diagram for a small time interval Δt

Therefore from Newton's 2nd Law, for a mass m moving in a circle of radius r, the centripetal force is given by:

$$F = m\frac{v^2}{r} = mr\omega^2$$

A centripetal force is not a force in its own right. It has to be *provided* by something. In the case of planets moving in a circular orbit around the Sun, the centripetal force between them is gravity.

Orbits and energy

The total energy (TE) a body has when it is in orbit is constant, and is the sum of its kinetic energy (KE) and potential energy (PE):

$$TE = KE + PE$$

For a mass m moving with velocity v the $KE = \frac{1}{2}mv^2$. The potential energy is the **gravitational potential energy** (GPE) and for two masses M and m separated by a distance r this is given by:

$$GPE = -G\frac{Mm}{r}$$

Note that the GPE varies as an inverse r relationship. The negative sign indicates that the GPE increases (gets less negative) with increasing distance. This is as you would expect – the higher an object is above the Earth's surface the greater gravitational potential energy it has. At an infinite distance ($r \to \infty$) the GPE tends to zero.

We may then write the total energy as

$$TE = \frac{1}{2}mv^2 - G\frac{Mm}{r}$$

Escape speed

By considering the energy of the orbit we can derive a simple expression for the speed a body needs to escape completely from the gravitational field of a large mass.

Geometrically, escaping from an orbit means making the orbit so elliptical that the major axis is infinitely long. Such a trajectory is called a **parabola** and the escape speed v_{esc} is the necessary speed needed to achieve this path of no return.

After leaving the Earth at escape speed, a body will have zero speed when it is just out of reach of the Earth's gravitational field. This will occur at a very large distance or, theoretically, when the body is at an infinite range from the Earth. At this point, the KE will be zero as will the GPE, which will tend to zero as r gets very large. Therefore, at an infinite distance, we conclude that the total energy TE must also be zero:

$$0 = \frac{1}{2}mv_{esc}^2 - G\frac{Mm}{r}$$

Rearranging, we get:

$$v_{esc} = \sqrt{\frac{2GM}{r}}$$

This equation gives the escape speed starting from an initial distance r from a mass M and tells us that the further you are away from a central body, the slower you have to travel in order to escape from the remainder of the gravitational field.

WORKED EXAMPLE 2.4

What is the escape velocity from the Earth's surface?

Taking the mass of the Earth as 6×10^{24} kg and its radius as 6.4×10^6 m:

$$v_{esc} = \sqrt{\frac{2 \times (6.67 \times 10^{-11} \text{ N m}^2 \text{ kg}^{-2}) \times (6 \times 10^{24} \text{ kg})}{6.4 \times 10^6 \text{ m}}} = 11.2 \text{ km s}^{-1}$$

Measuring the masses of stars and planets

We can use the Law of Gravitation and the equations of circular motion to obtain information about the mass of any astronomical object that has a satellite, if the distance between it and the satellite is known.

WORKED EXAMPLE 2.5

Assuming that the Earth moves round the Sun in a circular orbit of radius 1.5×10^{11} m, what is the mass of the Sun?

The centripetal force between the Earth and the Sun is gravity, so that:

$$G \frac{M_S M_E}{r_{ES}^2} = M_E \frac{v_E^2}{r_{ES}}$$

where r_{ES} is the radius of the Earth's orbit and v_E is its orbital velocity. Rearranging:

$$M_S = \frac{v_E^2 r_{ES}}{G}$$

and if T is the period of the Earth's orbit (1 year) then $v_E = 2\pi r_{ES}/T$ so that:

$$M_S = \frac{4\pi^2}{T^2} \cdot \frac{r_{ES}^3}{G}$$

1 year $= 3.0 \times 10^7$ s and $r_{ES} = 1.5 \times 10^{11}$ m so we obtain:

$$M_S = \frac{4\pi^2}{(3.0 \times 10^7)^2} \cdot \frac{(1.5 \times 10^{11} \text{ m})^3}{(6.67 \times 10^{-11} \text{ N m}^2 \text{ kg}^{-2})} = 2.0 \times 10^{30} \text{ kg}$$

The total energy equation enables us to make some general statements about orbital motion. When an orbiting object's total energy is *negative* then the orbit is *bounded*, in other words the object is confined to an elliptical orbit around a massive body. If the total energy is exactly *zero* then it moves on an escape path following a parabolic trajectory. If the total energy is *positive* then the orbit is entirely unbounded and it moves along a trajectory, the shape of which is termed *hyperbolic*. These three motions are illustrated in Figure 2.10.

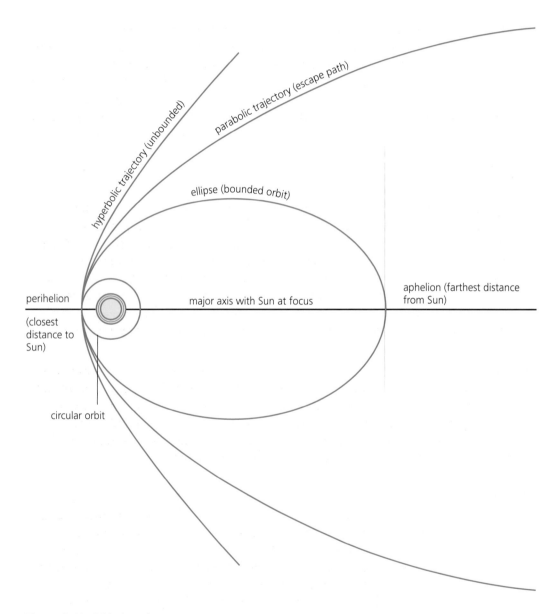

Figure 2.10 Orbital motion

Summary

◆ What we can learn from space comes to us mainly in the form of electromagnetic radiation which can be described either in terms of waves or particles. Particles of electromagnetic radiation are called **photons** and have an energy given by the **Planck relation**. Photon energies are characterised by their wavelength: short wavelengths mean high energy, long wavelengths mean low energy.

◆ Matter is made up of atoms consisting of protons, neutrons and electrons. The electrons exist in discrete **energy levels** in atoms. Photons are emitted or absorbed by an atom when an electron undergoes a **transition** from one energy level to another. The energy of the photon when emitted or absorbed is **quantised** and is equal to the difference in energy levels. A useful model of the hydrogen atom is the **Bohr model**, which can be used to explain the spectrum of atomic hydrogen. The **Balmer lines** of the hydrogen spectrum are important in classifying stars by their spectra. Electromagnetic radiation is also emitted when charges are accelerated in electric and magnetic fields.

◆ A body will emit electromagnetic radiation with an intensity and wavelength related to its temperature. A **blackbody** is a hypothetical object that absorbs all radiation incident on it. Stars are good approximations to blackbodies. **Stefan's Law** describes the relation between power emitted and temperature for a blackbody radiator. Blackbody radiators have a continuous intensity distribution called a **Planck** or **blackbody curve**. The wavelength of maximum intensity on the Planck curve is related to temperature of the body by **Wien's Law**. Stefan's Law and Wien's Law may be used to estimate the surface temperature of stars.

◆ The **Doppler effect** is the apparent shift in wavelength of a wave motion due to the relative motion between the source and the observer. The Doppler effect for light enables astrophysicists to determine the velocity and direction of distant stars. Light from a star that is seen to be shifted towards the red end of the electromagnetic spectrum means that the star is receding, whereas light that is blue-shifted means that the star is approaching the observer.

◆ **Astrodynamics** is a branch of dynamics which deals with the motions of bodies that move in space under the influence of gravity. Three important principles in astrodynamics are **Kepler's Laws of Planetary Motion, Newton's Laws of Motion** and Newton's **Universal Law of Gravitation**. A special case of orbital motion under gravity is the **two-body problem** which deals with the motions of two bodies orbiting one another. Given the relevant data, we can use astrodynamics to calculate the masses of astronomical bodies and predict their positions and velocities both in the past and in the future.

47

Questions

1 Calculate the wavelengths of photons with the following energies: **a** 4.0 keV, **b** 4.0 eV, **c** 0.04 eV, **d** 4.0×10^{-10} eV. For each, give the part of the electromagnetic spectrum in which they lie.

2 In a hydrogen atom, an electron makes a transition from the 2nd excited state $E_2 = -1.5$ eV to the 1st excited state $E_1 = -3.4$ eV. What is the frequency and wavelength of the emitted photon?

3 The surface temperature T of the Sun is 5800 K and its radius is 6.96×10^8 m. Calculate a value for its luminosity L given that

$$L = \sigma T^4 \times surface\ area$$

where $\sigma = 5.67 \times 10^{-8}$ W m^{-2} K^{-4}

ULEAC, 1996, part

4 **a** The star Antares in the constellation Scorpio is a red star, while our own Sun is yellow. With reference to the idea of a blackbody, explain which star has the higher surface temperature,
   b What is the wavelength corresponding to the maximum radiation intensity of a star whose surface temperature is 3000 K? Stars range in temperature from about 2500 to 35 000 K; what is the corresponding range in wavelength of their peak emission?

5 A sunspot is an area on the Sun that appears dark only because it is cooler with respect to the rest of the Sun's photosphere. If we assume that the radiated power per square metre from a sunspot is 10% of that from the photosphere, what is the temperature of the sunspot if the temperature of the photosphere is 5800 K? What is the sunspot's wavelength of peak emission?

6 A particular line in the spectrum of a star is found to have a wavelength of 600.8 nm compared to 600 nm as measured in a laboratory. What is the speed of the star? Is it moving towards or away from us?

7 The Sun rotates on its axis. If we observe light from two points at opposite ends of a diameter the Doppler shift in wavelength is about 0.008 nm at $\lambda = 600$ nm. What is the angular velocity ω of the Sun about its axis of rotation, given that its radius is 7.0×10^5 km?

8 The Sun's speed around our galaxy is 220 km s^{-1} and its distance from the galactic centre is 2.8×10^4 light years. If the Sun's orbit is a circular Keplerian one, what is the mass of the galaxy within the radius of the solar orbit?

9 **a** State Newton's Law of Universal Gravitation. What kind of force law does it obey?
   b Two stars form a binary system. At one point in their orbital motion they are 3.2 AU apart. If they have masses $1.9 M_\odot$ and $2.5 M_\odot$ (where $M_\odot$ is the solar mass), what gravitational force do they exert on each other at this distance?

QUESTIONS

10 a State Kepler's three Laws of Planetary Motion.
    b Using the third law, calculate in years the period of Jupiter's orbit (assume that the orbit of Jupiter is perfectly circular).
       (Average distance of Jupiter from the Sun $r = 5.20\,AU$; mass of Sun $M_\odot = 2 \times 10^{30}\,kg$; $1\,AU = 1.5 \times 10^8\,km$.)

11 a Show that Kepler's 3rd Law is homogeneous with respect to units.
    b Communications satellites are often 'parked' in geosynchronous orbits above the Earth. These are orbits where the satellite appears to remain fixed above the point on the Earth's surface. At what altitude above the Earth must these satellites be?

12 The Hubble Space Telescope (HST) is in a near circular orbit at an altitude of approximately 600 km above the Earth's surface. What is its orbital period? If the telescope has a mass of 11 000 kg, calculate the total minimum energy that was needed to place it in this orbit.

Astrophysical measurements

The methods used to detect and collect radiation from space range over all parts of the electromagnetic spectrum. Astronomers and astrophysicists have developed many different kinds of measuring instruments which extend our ability to 'see' into the furthermost reaches of space. Since the universe is so vast, a number of special units are used to measure distances. In this chapter we will look at a number of techniques that are employed in making astrophysical measurements and some of the units which are used.

Measuring the universe

The detection of radiation from space provides us with the only real clues as to the physical processes occurring within stars and other astronomical objects. Various methods have been developed to collect, detect and analyse this radiation so that we can learn more about the structure of the universe and develop theories as to its origin.

Astronomers and astrophysicists use a number of measurement units that measure sizes from the microscopic to the extragalactic. Units for small quantities are the SI ones generally found in atomic and nuclear physics but on astronomical scales, particularly for distances, it is convenient to use units based on astronomical observations as well.

We will begin by looking at some important astronomical units of length and how they are defined. We will then see how various techniques are applied to measure these very large distances and to detect the weak levels of radiation often encountered in making astrophysical measurements. Finally, we will look at the ways in which astrophysicists analyse and use the information gathered.

Units of astronomical distance

The astronomical unit (AU)

A natural starting point for astronomical distances is the distance from the Earth to the Sun. Since planets move in ellipses around the Sun, the 'average' distance is taken to be the semi-major axis of the Earth's orbit. This distance is called the **astronomical unit (AU)** and

$$1 \text{ AU} = 1.496 \times 10^{11} \text{ m}$$

The astronomical unit is only appropriate on interplanetary scales, as the distances to other stars are so great as to render it too small to be useful. For the Moon, the Sun and some of the nearer planets, astronomers can use radar ranging to determine their distances. Table 3.1 shows the relative distances of the planets from the Sun, in AU.

The parsec (pc)

For interstellar distances, we use a unit called the **parsec (pc)**. In order to understand how the parsec is defined we need to look at some trigonometry and what is meant by **parallax**.

Table 3.1 Distances of the planets from the Sun

Planet	Mean distance (semi-major axis)/AU
Mercury	0.38
Venus	0.72
Earth	1.00
Mars	1.52
Jupiter	5.20
Saturn	9.54
Uranus	19.18
Neptune	30.06
Pluto	39.44

To see what parallax is, look at a pencil held at arm's length. If you alternately open and close one eye several times then the pencil will appear to jump back and forth relative to fixed points in the background. The angle that the pencil makes with your eye as it shifts to and fro is called the **parallax angle**, and if you bring the pencil closer than arm's length then the parallax angle gets larger. Parallax also happens on an astronomical scale where the pencil corresponds to a nearby star, and the fixed points in the background correspond to distant stars which do not appear to change their positions as the Earth orbits the Sun.

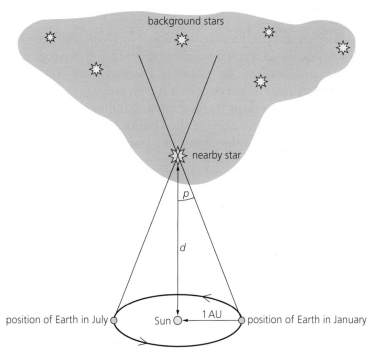

Figure 3.1 As the Earth orbits the Sun a nearby star will be observed to shift its position with respect to the distant background stars

Look at Figure 3.1. Suppose we record the position of a nearby star at two points on the Earth's orbit separated by a time interval of six months (note that these positions are separated by a distance of 2 AU). Owing to parallax, the nearby star appears to shift in position relative to the more distant stars. By using simple trigonometry we can show that the parallax angle p (measured in radians) is related to the distance d of the star by:

$$d = \frac{1\ AU}{\tan p} = \frac{1\ AU}{p} \quad (\text{using the fact that for small angles, } \tan p = p)$$

One radian = 57 degrees, 17 minutes, 44.81 seconds, and if we now measure all angular sizes in arcseconds (") then 1 rad = $(57 \times 3600)" + (17 \times 60)" + 44.81" = 206\ 264.81"$, or $1" = 1/(206\ 264.81)$ rad.

So $p"$ (in arcseconds) becomes $p/(206\ 264.81)$ radians and substituting for p in the equation above we get:

$$d = \frac{206\ 264.81}{p"}\ AU$$

We are now in a position to define the parsec. The word is an abbreviation of *parallax* and *second* and

$$1\ pc = 206\ 264.81\ AU$$

We can then write:

$$d = \frac{1}{p"}\ pc$$

This shows that when the parallax angle is equal to 1" the star's distance is 1 pc.

The final equation above is the fundamental definition: *a parsec is the distance at which the observed parallax of the star is equal to 1 second of arc.* You can now see why this unit is so useful. Once the parallax of a star in arcseconds is known then its distance is found by simply taking its reciprocal. Parallax that arises due to the motion of the Earth around the Sun is called **heliocentric parallax** and the measurement of the angular displacement of a star during half a year is called the method of **annual parallax**.

WORKED EXAMPLE 3.1

The measured annual parallax of the star *Proxima Centauri* is found to be 0.762". What is its distance in parsecs?

Using

$$d = \frac{1}{p"}\ pc$$

we find

$$d = \frac{1}{0.762} = 1.31\ pc$$

The distance to *Proxima Centauri* is therefore 1.31 pc.

Remember that 1 parsec = 206 264.81 AU = 206 264.81 $\times$ 1.496 $\times$ 10^{11} m = 3.1 $\times$ 10^{16} m. One thousand parsecs is one **kiloparsec (kpc)**; one million parsecs is one **megaparsec (Mpc)**.

The limit for measuring parallax angles using the diameter of the Earth's orbit is about 0.01 arcsecond, which means astronomers on Earth can measure distances to about 100 pc or 300 ly. With space astronomy, astronomers using the Hubble Space Telescope can make parallax measurements to a precision of 0.001" and the recently launched *HIPPARCOS* survey satellite is able to measure stellar parallaxes to 0.002".

The light year (ly)

One **light year (ly)** is the distance that a photon of light would travel through space in one year. Since light travels at 3 $\times$ 10^8 m s^{-1} we calculate this as 3 $\times$ 10^8 m s^{-1} $\times$ 365 days $\times$ 24 hrs $\times$ 3600 s = 9.46 $\times$ 10^{15} m.

Sub-units are the **light minute** = 3 $\times$ 10^8 m s^{-1} $\times$ 60 s = 1.8 $\times$ 10^{10} m, and the **light second** = 3 $\times$ 10^8 m s^{-1} $\times$ 1 s = 3 $\times$ 10^8 m.

$$1 \text{ parsec} = 3.262 \text{ ly}$$

Stellar magnitudes

When you look up at the stars at night, it is obvious that some stars are brighter than others. However, this is deceptive since the observed brightness of an object clearly depends on how far away you are from it. A 60 W light bulb at a distance of 7.5 m and a 100 W light bulb at 10 m both have the same *apparent brightness*, but put side by side the 100 W bulb is the more luminous of the two. Light as it spreads out from simple objects obeys an *inverse square law*. For example, a 100 W bulb appears 1/100 times dimmer at a distance 10 times further away.

The **luminosity** L of an object is the amount of energy in joules it radiates per second (i.e. its power) and is measured in watts. Just like the bulb, light emitted from a star spreads out as an inverse square law and if we imagine the star as a 'point source' centred on a sphere of radius R, then the energy passing through each square metre every second is simply the luminosity divided by the surface area of the sphere (Figure 3.2). We therefore define the **brightness** b of a star as:

$$b = \frac{L}{4\pi R^2} \text{ W m}^{-2}$$

and this represents the *light flux* received.

Astrophysicists prefer to talk about a star's brightness rather than its luminosity, and this is expressed in a scale of **magnitude**. The magnitude scale on which stars are rated is based on a convention first devised by Hipparchus in the second century BC. They are classified by *apparent* magnitude, i.e. how bright the star appears if you view it directly with your eye without any magnification. On this scale, the brightest stars seen with the naked eye have magnitude +1.0 and the faintest have magnitude +6.0.

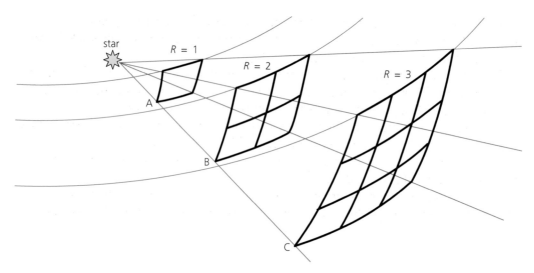

Figure 3.2 Brightness and the inverse square law. The luminosity of a star must illuminate an ever-increasing area of a sphere as the distance from the star increases. The brightness decreases as the square of the distance, i.e. the brightness of the star is nine times less at C than it is at A

With the invention of the telescope and for consistency, it later became necessary to assign to some stars a magnitude brighter than +1.0. Hence *Vega* has magnitude 0.0 and *Sirius* magnitude −1.4 whereas the Sun has a magnitude of −26.74. The apparent magnitude scale applies to any bright astronomical object. The Full Moon has magnitude −12.7 and Venus reaches a maximum apparent magnitude of −2.6.

The faintest object visible in the largest telescopes has apparent magnitude +21.0 and the faintest object that can be detected using the most advanced methods of light detection is +28. Table 3.2 shows the apparent magnitudes of some well known stars in descending order of brightness.

Table 3.2 Apparent magnitudes of some stars

Star	Apparent magnitude	Distance from the Earth/ly
Sun	−26.74	—
Sirius	−1.46	8.6
Canopus	−0.72	74
Arcturus	−0.097	34
Rigel	0.12	1400
Altair	0.77	16
Aldebaran	0.85*	60
Deneb	1.25	1500
Barnard's Star	9.5	6.0

*variable star

It is important to remember that the *more negative* the value of apparent magnitude, the *brighter* the star appears. Conversely, the *larger the magnitude* (more positive), the *fainter* the star appears.

The human eye perceives equal ratios of intensity of light as equal differences in brightness. In this way it is rather like the human ear which hears equal ratios of sound intensity as equal differences in loudness. On Hipparchus's scale, the brightness of stars of magnitude $+1.0$ was about 100 times greater than that of stars of magnitude $+6.0$. This means a difference of $6 - 1 = 5$ magnitudes corresponds to a brightness ratio of 100. A magnitude difference of 1 therefore corresponds to a brightness ratio of $(100)^{1/5}$ or 2.512.

An English astronomer Norman Pogson (1829–91), in 1856, formulated Hipparchus's somewhat subjective scale into a precise mathematical law expressed as:

$$m_2 - m_1 = -2.5 \log \left(\frac{b_2}{b_1} \right)$$

where:

m_1 = apparent magnitude of star 1
m_2 = apparent magnitude of star 2
b_1 = received brightness of star 1
b_2 = received brightness of star 2

We can see that Pogson's Law is consistent if we note that the minus sign ensures that magnitudes are a measure of faintness. If b_2 is less than b_1 (star 2 fainter than star 1) then the brightness ratio is less than unity and the log of the ratio is negative. So $m_2 - m_1$ is positive and the faintest star has the larger magnitude, as expected. The multiplier 2.5 is simply a scaling factor that ensures that a brightness ratio of 100 corresponds to a magnitude difference of 5.

Absolute magnitudes

Now suppose you could place all the stars at a fixed distance from the Earth. The differing distances would not then be a factor in how bright the stars appeared. Instead, the differences in magnitude would be due *only* to differences in luminosities and in this case the magnitude of a star would be its *absolute* magnitude.

Astronomers use a standard distance of 10 parsecs for absolute magnitude comparison. We therefore define the magnitude that a star would have if it was placed 10 pc from the Earth as its **absolute magnitude**. Once again, absolute magnitudes operate on the magnitude scale with negative values being brighter.

What then is the relationship between a star's apparent magnitude m (which we see) and its absolute magnitude M?

If a star has luminosity L, then assuming that its luminosity is radiated uniformly over the area of a sphere, let the brightness received at the Earth if the star is at a distance d be b_d. and let b_{10} be the brightness the star would have at a distance of 10 pc. Then using Pogson's Law we find that:

$$m - M = -2.5 \log\left(\frac{b_d}{b_{10}}\right)$$

but the brightness b_d is equal to $L/4\pi d^2$, so that

$$\frac{b_d}{b_{10}} = \frac{L}{4\pi d^2} \div \frac{L}{4\pi(10)^2} = \left(\frac{10}{d}\right)^2$$

where d is expressed in parsecs. Therefore:

$$m - M = -2.5 \log\left(\frac{10}{d}\right)^2$$

or, using the properties of logarithms,

$$m - M = -5(\log 10 - \log d)$$
$$= 5 \log d - 5$$

and, rearranging, we can express the distance as

$$d = 10^{(m-M+5)/5} \text{ pc}$$

The above equation relates apparent magnitude m, absolute magnitude M and distance from the Earth d in parsecs. If a star's distance is known and its apparent magnitude measured, then we can determine its absolute magnitude.

WORKED EXAMPLE 3.2

The star *Procyon A* has an apparent magnitude of +0.4 and lies at a distance of 3.5 pc from the Earth.

What is its absolute magnitude?

Using $M = m - 5 \log d + 5$

$$M = (+0.4) - 5 \log 3.5 + 5$$
$$= +2.7$$

Procyon A has an absolute magnitude of +2.7.

Bolometric magnitudes

In practice, however, it's not quite as simple as this. Allowances have to be made for the fact that some stars may emit a significant amount of their light in the non-visible part of the electromagnetic spectrum. For example, very luminous hot stars appear dim to the eye only because they emit most of their radiation in the ultraviolet and we must remember that the Earth's atmosphere absorbs many non-visible wavelengths. It follows that absolute magnitudes as deduced from apparent magnitudes using this distance–magnitude relation need to be corrected for these effects. To do this, astronomers define a **bolometric magnitude** which is the star's apparent magnitude measured *above* the Earth's atmosphere over all wavelengths of electromagnetic radiation and is determined by an instrument called a **bolometer** (see page 74). For the Sun, which emits most of its radiation in the visible part of the electromagnetic spectrum, the bolometric correction is nearly zero.

It is usually made clear which type of magnitude is being used in magnitude measurements. Astronomers refer to the **visual magnitude** as the star's magnitude over the wavelength range of the human eye, while the **bolometric magnitude** is that measured at all wavelengths.

Luminosity and magnitude

Luminosities are normally expressed in terms of the Sun's luminosity, which is termed $L_\odot$. For two stars 1 and 2 situated at the *same distance d*, the ratio of their brightnesses is equal to the ratio of their luminosities, i.e.

$$\frac{b_2}{b_1} = \frac{L_2}{4\pi d^2} \div \frac{L_1}{4\pi d^2} = \frac{L_2}{L_1}$$

and using Pogson's Law we can write:

$$m - M = -2.5 \log\left(\frac{b_2}{b_1}\right) = 2.5 \log\left(\frac{L_2}{L_1}\right)$$

Now suppose we let star 1 be the Sun and we use Pogson's Law to express the difference in *absolute magnitude* of star 2 and the *absolute magnitude* of the Sun. We can then write:

$$M_2 - M_{Sun} = -2.5 \log\left(\frac{L_2}{L_\odot}\right)$$

What is the value of M_{Sun}? The Sun is at a distance of 1 AU or 4.848×10^{-6} pc and has an apparent magnitude of -26.8. Using Pogson's Law,

$$M_{Sun} = m_{Sun} - 5 \log d + 5$$
$$= (-26.8) - 5 \log(4.848 \times 10^{-6}) + 5 = +4.77$$

The absolute magnitude of the Sun is $+4.77$.

We can now express the relationship between a star's absolute magnitude and its luminosity in terms of the Sun's luminosity as:

$$M = 4.77 - 2.5 \log\left(\frac{L}{L_\odot}\right)$$

(It is also a simple matter to show that the relationship between the star's *apparent* magnitude m and its brightness b is $m = 4.77 - 2.5 \log(b/b_{10\odot})$ where $b_{10\odot}$ is the brightness of the Sun at a distance of 10 pc.)

We now have an equation relating a star's luminosity to its magnitude and so if we know the star's absolute magnitude then we can work out how much power the star is radiating from its surface.

WORKED EXAMPLE 3.3

The star *Sirius* has an apparent visual magnitude of -1.46 and lies at a distance of 2.64 pc. How luminous is it compared with the Sun?

First of all we calculate the absolute visual magnitude of *Sirius* using Pogson's Law:

$$\begin{aligned} M &= m - 5 \log d + 5 \\ &= -1.46 - 5 \log(2.64) + 5 \\ &= +1.4 \end{aligned}$$

We can now calculate the luminosity of *Sirius* using $M = 4.77 - 2.5 \log(L/L_\odot)$:

$$1.4 = 4.77 - 2.5 \log\left(\frac{L_{\text{Sirius}}}{L_\odot}\right)$$

so

$$\log\left(\frac{L_{\text{Sirius}}}{L_\odot}\right) = 1.35$$

$$\therefore L_{\text{Sirius}} = L_\odot \times 10^{1.35} = 22.4 \, L_\odot$$

So *Sirius* is at least 22 times as luminous as the Sun.

We can now see how important measuring stellar magnitudes is if we want to find out the distance and luminosities of stars. Don't worry too much about the derivations but make sure you know what the individual terms mean and how they are used. The magnitude relations are summarised for you at the end of this chapter.

Units and conventions

In our discussion of magnitude measurements we have seen that astrophysicists express luminosities of stars as multiples of the Sun's luminosity $L_\odot$ ($= 3.90 \times 10^{26}$ W). This is also true of other quantities so that the mass of the Sun $M_\odot$ ($= 1.989 \times 10^{30}$ kg) and the radius of the Sun $R_\odot$ ($= 6.96 \times 10^8$ m) can be used for comparison, with the subscript $\odot$ as the accepted symbol for the Sun. For example, the luminosity of the star *Sirius* A is 22 $L_\odot$, its mass is 2.3 $M_\odot$ and it has a radius of 1.8 $R_\odot$.

A unit used by radio astronomers is the **Jansky**. The amount of energy received from astronomical objects is generally very small and at radio wavelengths the 'radio brightness' is measured per unit frequency. The Jansky, named after Karl Jansky (1905–50) who first discovered radio waves from space, is defined as 1 Jansky (Jy) $= 10^{-26}$ W m^{-2} Hz^{-1}.

Radio emissions from the Sun during intense solar activity are as high as 10^8–10^9 Jy although most celestial sources emit less than a few Jy.

Astronomical nomenclature

When discussing objects like stars, galaxies and nebulae, astronomers commonly use constellations to indicate their locations for reference purposes. Many of the brighter stars have proper names, like *Sirius* or *Canopus*. The letters of the Greek alphabet are assigned to the more conspicuous stars in successive order of decreasing brightness. For example, the brightest star in the constellation *Cygnus* (called *Deneb*) is named α *Cygni*, while *Denebola*, the second brightest star in the constellation *Leo* is called β *Leonis*.

Many star atlases and catalogues have been produced. Of note is the Messier catalogue prepared in 1781 by the French astronomer Charles Messier (1730–1817). Messier objects have the letter M followed by a number. M31, for example, is the great spiral galaxy in the constellation of *Andromeda*. Many observatories have produced their own catalogues for specific classes of objects or for those in certain regions of the sky, but the majority of stars that are both too faint and too numerous to catalogue remain nameless.

Certain symbols are also used for objects in the Solar System. These astronomical symbols are listed below:

	✱ Star	
☉ Sun	⊕♁ Earth	♅ Uranus
☾ Moon	♂ Mars	♆ Neptune
☿ Mercury	♃ Jupiter	♇ Pluto
♀ Venus	♄ Saturn	☄ Comet

Measuring astronomical distances

How does one begin to measure the distance to the stars and galaxies?

The starting point is the method of trigonometrical parallax as described in our discussion of the parsec (page 50). If the parallax angle of a star is known, then its distance can be calculated by trigonometry. However, the parallax angles are very small and cannot be measured from the ground to accuracies of better than 0.01" This only enables measurement of distances up to 100 pc.

Other parallax methods involve the motion of the Sun among the nearby stars. The Sun moves relative to the fixed stars at a velocity of 20 km s^{-1} (or about 4.1 AU per year), in the direction of the constellation *Hercules*. In ten years the Sun moves through a distance of about 40 AU, giving us a baseline that would theoretically enable us to measure distances to 2000 pc.

We must remember however that the stars do not remain stationary in space but are all moving with their angular displacements or **proper motions** across the sky just like the Sun, so we have to look at a group of stars and calculate the mean parallax of the group. The way this is done is somewhat complex and involves intricate statistical calculations, but the result is that we can reliably measure distances to a few hundred parsecs using this **statistical parallax** method.

For distances from a few hundred to a thousand pc astronomers use the method of **spectroscopic parallax**. This term is a misnomer since it is not a parallax method at all. Spectroscopic parallax involves using the spectrum of a star in order to locate its position on the Hertzsprung–Russell Diagram (see Chapter 5). This is a diagram that shows the absolute magnitude of a star in relation to its spectral type. Once the spectral type has been identified, then from the Hertzsprung–Russell Diagram the star's absolute magnitude can be determined. Knowing the star's apparent magnitude by observation, Pogson's Law can then be used to calculate its distance in pc.

The period–luminosity (P–L) relation

Beyond a few thousand parsecs astronomers use distance indicators to find out how far away an object is. In Chapter 1 (page 14) we mentioned the existence of variable stars called Cepheids whose luminosity varies in a periodic fashion with time. It is possible to use them to estimate distances.

The star *Delta Cephei* is one such star and its variation in luminosity with time has been extensively studied. The light curve of *Delta Cephei* is shown in Figure 3.3. Astronomers have found many other stars that behave like *Delta Cephei* and they all go under the collective name of **Cepheids**. The interesting feature about Cepheids is the relationship between their average luminosities and periods, namely that *the longer the period, the more luminous the Cepheid star is.*

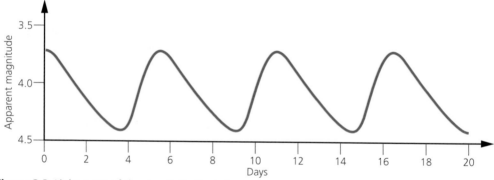

Figure 3.3 Light curve of the star *Delta Cephei*

Figure 3.4 is a diagram that shows the relationship between period and luminosity of Cepheids, as a function of absolute magnitude with time in days. There are in fact two types of Cepheids, I and II, which differ only in spectral characteristics and the shape of their light curves, but their period–luminosity relation is of the same form. Also included on the graph is a third type of variable star called *RR Lyrae* stars, which are found mainly in globular clusters and differ from type I and II Cepheids by having shorter periods. The P–L relation was calibrated from apparent to absolute magnitudes by measuring the distance to a single Cepheid by statistical parallax methods. This was done for one of the nearer Cepheids, the star *Polaris*, which at a distance of 200 pc varies in intensity by about 0.03 magnitudes over a period of 3.97 days.

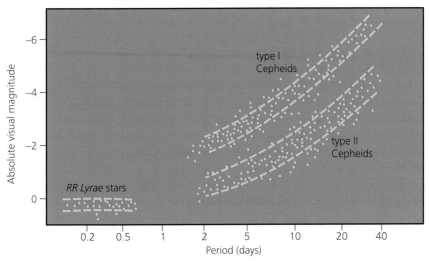

Figure 3.4 The period–luminosity diagram

We can use the period–luminosity relation to find the distance to a Cepheid variable star by using the following steps:

Step 1 Locate a Cepheid (you can do this be seeing if it has a periodic light curve).
Step 2 Measure its period by observing it nightly.
Step 3 Find the star's *absolute* magnitude using the $P-L$ relation.
Step 4 Measure the star's received light flux to determine its *apparent* magnitude (this is done with an instrument called a photometer (see page 77).
Step 5 Calculate the star's distance in pc using $d = 10^{(m-M+5)/5}$ pc.

WORKED EXAMPLE 3.4

A certain variable star is found to have a period of 5 days and an apparent visual magnitude of +8. How far away is it?

From the $P-L$ relation (Figure 3.4) a period of 5 days corresponds to an absolute visual magnitude of -2.7.

Using the distance–magnitude relation $m - M = 5 \log d - 5$ we obtain:

$$\log d = \frac{(+8) - (-2.7) + 5}{5} = 3.14$$

$$\therefore d = 1380 \text{ pc}$$

The star is 1380 parsecs distant.

The advantage of using variable stars as distance indicators is that only two measurements are needed – the average apparent magnitude and the period. However, an assumption made is that there is no absorption of the starlight by interstellar matter, which is an important consideration and is discussed in

61

Chapter 5. Variable stars that are used in this way to estimate distances are an example of what are called **standard candles** – objects that can be used as 'yardsticks' to determine astronomical distances to stars and galaxies.

The period–luminosity relation has enabled us to establish firmly the size of our own galaxy, but what about the distances to other galaxies and indeed the furthermost reaches of space? For nearby galaxies in the Local Group, we can observe Cepheids in them and use the $P-L$ relation to estimate their distances, but to estimate the size of the universe as a whole, or the **cosmological distance scale**, requires a combination of other methods which are discussed in Chapter 10.

Ground-based observing techniques

Since, in most cases, the radiation received from celestial objects is weak, collectors of radiation are needed so that enough energy can be gathered for measurements to be made. Also, since the Earth's atmosphere is only transparent to certain parts of the electromagnetic spectrum (see Figure 2.2), we are limited in wavelength as to what we can see from the ground. Some observations can therefore only be made above the atmosphere using space astronomy; the type of instrument used depends on the wavelength of the radiation.

Most visual (optical) and radio astronomy can be done from the ground. Once the energy is collected it must be analysed and detected so that information can be extracted from it. In this section we will briefly look at some collectors of celestial radiation that are used in ground-based astronomy.

Optical telescopes

The angular size of an object depends on the angle that it subtends at the eye (Figure 3.5). The further away something is then the smaller the angle subtended. The invention of the optical telescope revolutionised our knowledge of astronomy by making distant objects appear closer and therefore larger. There are many different designs of telescope but all use an **objective**, either in the form of a lens or a reflecting mirror, to concentrate starlight at a **focus**. An eyepiece lens then acts like a magnifying glass to magnify this focused image.

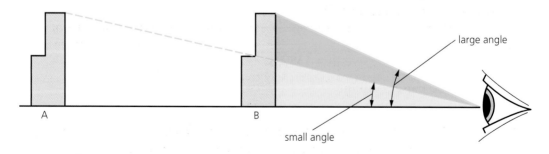

Figure 3.5 The apparent size of something depends on the angle it subtends at the observer's eye. The building at A appears smaller compared to the one at B because it subtends a smaller angle at the eye

Three important features of a telescope are its:

1 magnification
2 light gathering power
3 image resolution

Figure 3.6 shows the principle of a simple **refracting telescope**. Light from a distant star enters the objective lens and an image is formed at the focus F. This image acts as an object for the eyepiece whose focal point a distance f from the lens coincides with the position of F. A greatly magnified image is thus formed at infinity. The magnifying power of a telescope comes about due to the fact that the angle subtended at the eye by the image formed by the eyepiece is greater than that formed by the distant star with the naked eye.

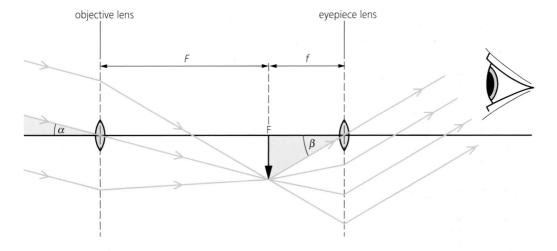

Figure 3.6 A refracting telescope. The angle β subtended at the eye by the final image is greater than the angle α subtended at the eye by the object without the telescope

There are several different designs of **reflecting telescope** but all use an objective mirror to bring light from a distant object to a focus F, where an eyepiece can be used to magnify the image (Figure 3.7).

The Hubble Space Telescope (page 70) incorporates a reflecting telescope of Cassegrain focus design. The **magnification** of a telescope is found by the ratio:

$$\frac{\text{focal length of objective}}{\text{focal length of eyepiece}} \quad \text{or} \quad \frac{F}{f}$$

(In the case of the simple refractor the image is upside down, but this does not matter for astronomical use.) In practice, astronomers would employ a range of eyepieces with f say from 9 mm to 40 mm, in order to view using a range of magnifications for a fixed F.

(a) Prime focus

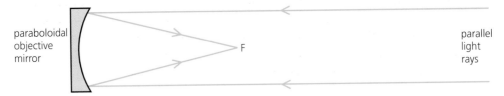

(b) Newtonian focus

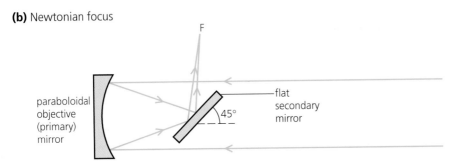

(c) Cassegrain focus

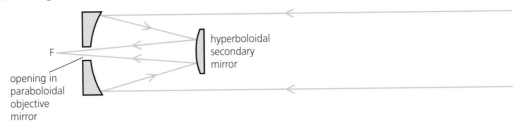

Figure 3.7 Three designs of reflecting telescope

A telescope is rather like a bucket for collecting photons. Astronomers like big telescopes because of their ability to gather light, and a measure of how much light one can collect is called the **light gathering power (LGP)**. The LGP of a telescope is directly proportional to the square of the diameter of its objective. This is because the surface area of a circular object of diameter a is equal to $\pi \times (\frac{1}{2}a)^2 = \frac{1}{4}\pi a^2$. The LGP is a relative measure for comparing the ability of different telescopes to 'grasp' light.

WORKED EXAMPLE 3.5

Compare the LGP of a reflecting telescope whose primary (objective) mirror diameter is 0.5 m with that of a refracting telescope with an objective lens of diameter 60 mm.

The ratio of the respective LGPs is $\frac{1}{4}\pi(0.5)^2 \div \frac{1}{4}\pi(0.06)^2 = 69.4$.

The 0.5 m telescope has an LGP 69 times as great as the 60 mm one.

The third measure of a telescope's performance is its **image resolution**. This is measured in terms of its angular resolving power, which is the minimum angle between two objects that can be clearly seen to be separated. Image resolution depends on the wavelength of the light and on the diameter of the objective, or **aperture**. It can be shown that, for a circular objective of diameter a, the minimum resolvable angle θ_{min} for light of wavelength λ is given by:

$$\theta_{min} = 1.22 \frac{\lambda}{a} \text{ radians}$$

or, since 1 radian = 206 265 arcseconds,

$$\theta_{min} = 2.52 \times 10^5 \frac{\lambda}{a} \text{ arcseconds}$$

WORKED EXAMPLE 3.6

A telescope of aperture 0.3 m is observing a star emitting light of wavelength 510 nm. What is the image resolution in radians?

$$\theta_{min} = 1.22 \times \frac{510 \times 10^{-9}}{0.3} = 2.1 \times 10^{-6} \text{ rad}$$

Table 3.3 lists some large ground-based telescopes currently in use.

Table 3.3 Ground-based optical telescopes

Name	Location	Aperture/m	Focal length/m
W.M. Keck	Hawaii, USA	10.0	17.5
George E. Hale	Mt Palomar, USA	5.08	16.8
Multiple Mirror Telescope (MMT)	Mt Hopkins, USA	4.5	4.9
CTIO	Cerro Tolodo, Chile	0.4	11.2
Anglo-Australian Telescope (AAT)	Australia	3.89	12.7
William Herschel Telescope (AAT)	La Palma, Canary Islands	4.2	10.5
UK Infrared Telescope (UKIRT)	Hawaii, USA	3.8	9.5
Keck Telescope [†]	Hawaii, USA	10.0	17.5
Gemini North	Hawaii, USA	8.1	129.6*
Gemini South (operational 2001)	Cerro Pachon, Chile	8.1	129.6*

[†]The Keck Telescope uses an assembly of small segmented mirrors to form a large composite objective mirror.

*The Gemini telescopes use a Cassegrain focus system that 'folds' the focal length up to make the telescope more compact.

Limitations of optical telescopes

Although the atmosphere readily admits electromagnetic radiation at optical wavelengths, in practice the 'seeing' is limited by atmospheric conditions, particularly by convection currents and air turbulence, as well as by minute changes in the primary mirror shape due to mechanical stresses. Two techniques, **active optics** and **adaptive optics**, have been devised to overcome these limitations for reflecting telescopes.

With active optics, alterations in the mirror shape due to mechanical stresses are compensated every few minutes by a computer that controls mechanical actuators attached to the mirror. This ensures that the image is always kept in the sharpest focus.

Adaptive optics is concerned with compensating for the effects of atmospheric turbulence. The optical wavefronts from a reference star are distorted as they travel through the atmosphere. These distortions are measured and the information sent by a computer to piezoelectric transducers that distort the mirror shape so that it 'matches' the distortion due to the turbulent atmosphere. The principle is to cancel out the effect of the turbulence, resulting in better image resolution. The mirror shape is continually distorted over a timescale of 10–100 ms. In order to reduce structural stresses, a smaller secondary mirror in the telescope's optics is distorted rather than the much larger primary mirror. The twin Gemini telescopes represent a new generation of telescope that use both active and adaptive optics as an integral part of their advanced design.

Advantages of reflecting telescopes

One of the reasons why reflecting telescopes are frequently used is because they are suitable for the compensating techniques discussed above. Other advantages are:

1 Mirrors can be made lighter for the same aperture (objective lenses in big telescopes are very heavy and prone to structural distortion).
2 Mirrors can be more easily manufactured in large sizes.

Radio telescopes

The science of radio astronomy was born when the telephone engineer Karl Jansky (see page 58) was looking for sources of static noise affecting radiotelephone communications. Using a radio aerial, Jansky found that some of this noise was coming from the Milky Way. Later, other workers found numerous astronomical radio sources including the Sun, and **radio telescopes** were developed to study them.

The simplest radio telescope consists of a 'flux bucket' in which radio energy is collected in a parabolic dish aerial and brought to a focus at a receiver (Figure 3.8a). A radio telescope has a low resolving power because of the long wavelengths of radio waves. (Remember that the angular resolving power is proportional to the wavelength and that radio waves are electromagnetic waves just like light except that their wavelengths are typically 100 000 times longer.) It follows that for an optical telescope and a radio dish of the same aperture, the radio telescope will have at least 100 000 times *lower* resolution.

WORKED EXAMPLE 3.7

Compare the theoretical image resolution of the Hale telescope on Mount Palomar with that of a radio telescope of the same aperture working at a wavelength of 2 m.

From Table 3.3, the aperture of the Hale telescope is approximately 5.1 m. Taking the wavelength of light to be 510 nm,

$$\text{image resolution} = 1.22 \times \frac{510 \times 10^{-9}}{5.1} = 1.2 \times 10^{-7}\,\text{rad}$$

For the radio telescope,

$$\text{image resolution} = 1.22\,\frac{2}{5.1} = 0.5\,\text{rad}$$

The radio telescope has an image resolution of some 4×10^6 less than an optical one.

Clearly we can improve resolution by increasing the size of the parabolic dish but this is impractical. To build a radio telescope working at 2 m with a resolution the same as the Hale optical telescope would require a dish with a diameter about 20 000 km! However, there is a way to increase the resolving power without having to build huge dishes. Radio astronomers use **interferometry** to improve the image resolution.

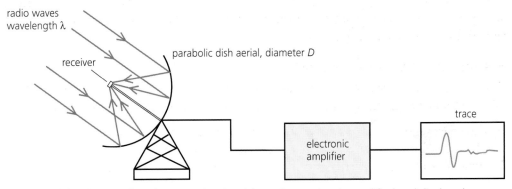

Figure 3.8 (a) A basic radio telescope. The signal from the receiver is amplified and displayed as an intensity trace. The resolution is given by 1.22λ/D

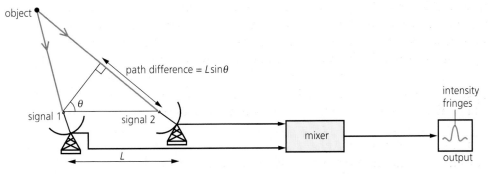

Figure 3.8 (b) Principle of a radio interferometer. If the path difference is a whole number of wavelengths then the two signals constructively interfere. The resolution of this system is 1.22λ/L

To understand how a radio interferometer works, look at Figure 3.8b. Two identical dish aerials are placed a distance L apart and their signals fed into a receiver which mixes them together. If a celestial source is directly overhead then the signals from it arrive at the aerials in phase and constructively interfere, giving a strong signal. Conversely, when the signals are 180° out of phase the signals destructively interfere. As the source moves across the sky an interference pattern of maxima and minima is recorded, exactly like that for light passing through a double slit. The angular distance between successive maxima is the resolving power of the radio interferometer and it can be shown that this equals $1.22\lambda/L$.

So if we make L very large then we can increase image resolution; an interferometer has an image resolution equal to that of a single dish aerial of diameter L. The interferometer 'sees' the source as a series of maxima and minima along a narrow strip of sky in the direction of the baseline. In order to build up a complete image of the source we need to have interferometers at different orientations. The Very Large Array in New Mexico (Figure 3.9) is a radio interferometer which together with the rotation of the Earth is able to map the sky at radio wavelengths. This technique of mapping a complete object is called **aperture synthesis** and involves sophisticated signal processing.

Figure 3.9 The Very Large Array radio interferometer in New Mexico, USA

Even better resolution can be obtained by Very Large Baseline Interferometry (VLBI). In VLBI, the signals from a common source are received by radio telescopes that are very long distances apart, located for example on different continents. The signals are recorded on magnetic tape and stored in a computer. If the time of observation and the locations are accurately known then the signals can be combined to give an image. The longest baseline possible by this method is the diameter of the Earth, giving $L = 12\ 000$ km.

Infrared astronomy

Observing from the ground in the infrared range is limited to only a few wavelengths due to absorption by carbon dioxide and water vapour in the Earth's atmosphere. An optical telescope is used with a bolometer (see page 74) placed at its focus. The telescope concentrates the infrared energy onto the bolometer which measures the increase in temperature due to the electromagnetic radiation incident on it. The bolometer produces a voltage output from which the temperature of the radiation can be determined. Because the range of infrared wavelengths that can be observed from the ground is so limited, much of infrared astronomy is done using space astronomy; IRAS (*Infra Red Astronomy Satellite*) is a notable example (see Figure 3.10).

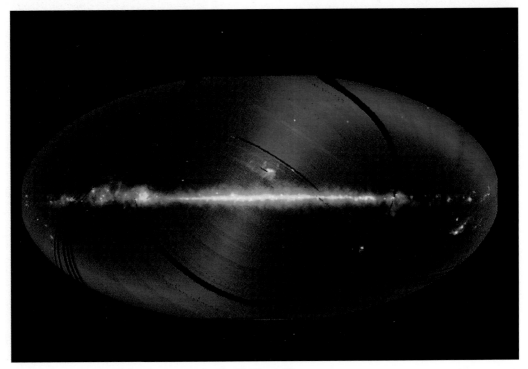

Figure 3.10 A map of the sky as seen by the IRAS satellite

Space astronomy

The use of high altitude balloons, rockets, satellites and spacecraft has meant that we can observe space at wavelengths that would normally be absorbed by the Earth's atmosphere. This method of observing the universe is commonly called **space astronomy**.

We have already mentioned IRAS, which surveyed the sky over a range of infrared wavelengths using a telescope of aperture 57 cm and mapped over 200 000 infrared sources. It also provided valuable information relevant to processes involving the formation of stars.

Figure 3.11 The Hubble Space Telescope in the cargo bay of the Space Shuttle *Endeavour*

The most famous space telescope is the Hubble Space Telescope (HST) (Figure 3.11). Despite initial problems with the primary mirror which were later corrected by shuttle astronauts, the HST has regularly returned spectacular images of a wide range of celestial objects and made important contributions to astrophysics. It consists of a telescope with a primary mirror diameter of 2.4 m and carries two main instruments, the Wide Field Planetary Camera (WFPC) and the Faint Object Camera (FOC), and is able to observe over the wavelength range from 110 nm to 1.1 μm.

X-ray astronomy

A rocket-borne experiment launched in 1962 provided the first direct evidence that some celestial objects emit **X-rays**. Detectors aboard the rocket revealed a sharp increase in X-ray photons as they were pointed in the direction of the constellation *Scorpius* and also in the direction of the constellations of *Sagittarius* and *Cygnus*. In addition, the photon count never dropped to zero, indicating that the sky was immersed in an X-ray background of interstellar origin. Subsequent space missions have located many more X-ray sources, revealing information about the number and arrangement of X-ray sources in the universe. X-ray sources are of great interest to astrophysicists as they are associated with highly energetic processes involving very high temperatures.

Up to certain energies, it is possible, by using specially constructed mirrors, to focus X-rays by reflection to form images. Such a system is called a **grazing incidence telescope**, as reflection of X-rays only occurs at angles of incidence approaching 90°. An X-ray telescope is a grazing incidence instrument which embodies a highly polished parabolic surface followed by a hyperbolic one. The mirror surfaces are coated with a thin film of gold so that the X-ray photons are reflected and not absorbed. Image resolution of a few arcseconds is regularly achievable with this instrument, and the collecting area of a practical telescope is maximised by 'nesting' a series of mirror combinations inside each other.

Gamma ray astronomy

During the Cold War, the US military launched a series of satellites called *Vela* that were designed to ensure compliance with the 1963 Nuclear Test Ban Treaty. The *Vela* satellites were able to detect the flash of gamma rays given off by the detonation of a nuclear weapon and they did indeed detect gamma ray flashes; however, they were not coming from the direction of the Earth, but from outer space! Astrophysicists now call these events **gamma ray bursters (GRBs)**. They last from about 0.01 to 1000 s with energies ranging between 1 keV and 100 MeV. In 1991 the space shuttle *Atlantis* released the Compton Gamma Ray Observatory (CGRO). This satellite is equipped with sensitive gamma ray detectors (see page 73) that can assign position information to the bursts in order to correlate them with known objects. So far the CGRO has found that the sources are spread across space but their distribution is not uniform. The origin of GRBs is currently a mystery and it is not known whether they originate within our galaxy or outside it.

Detectors and imaging devices

Detecting and recording electromagnetic radiation from space has advanced far from the time when drawings were first made by astronomers looking through the eyepiece of a telescope. Modern detection technology has extended the range of wavelengths and intensities normally accessible to the human eye and the processing of images by computers can reveal information and structure not immediately discernible to the brain. These days an astronomer may be found in a warm room watching an image on a TV screen – he or she may be located hundreds of miles from the telescope controlling it by remote communication links!

An important measure of detector's sensitivity to electromagnetic radiation is its **quantum efficiency (QE)**. This is defined as:

$$QE = \frac{\text{number of photons detected}}{\text{number of photons incident}} \times 100\%$$

The quantum efficiency tells us how well a detector can capture photons and make them available for further amplification. The eye has a very low QE of about 1% (about one photon in every hundred is detected). An ideal detector would have a QE of 100%; detectors used for astronomical work have values ranging from about 4% up to 80% or more.

Photographic emulsion

Photography is extensively employed to record astronomical images. A photographic emulsion is made up of numerous tiny grains of silver bromide a few μm across. The chemical development of these grains is activated by incident photons, forming a **latent image**. The grain size determines the image resolution. The emulsion is coated onto a rigid glass plate from which positions and intensities of stars can be accurately measured. Photographic plates have long storage lifetimes and emulsions can also be optimised for certain wavelengths, with the resulting image digitised for input into a computer. The QE of photographic emulsion is quite low, being only a few per cent. However, it is relatively cheap, simple to use and can compensate for its low QE if exposures are made over long time periods. In addition, the photographic emulsion presents a large sensitive surface area to the image.

Photomultiplier tubes (PMTs)

These make use of the **photoelectric effect**. An incident photon striking the surface of certain metals will eject electrons from it with energy given by:

$$E(\text{photoelectron}) = hf - \Phi$$

where f is the frequency of the incident photon (and hf its energy) and Φ is the **work function**, which is the energy needed to remove an electron from the surface of the metal.

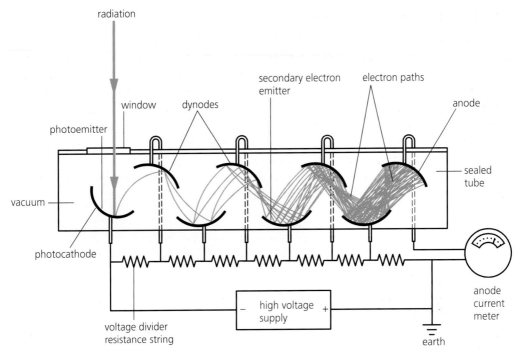

Figure 3.12 A photomultiplier tube (simplified arrangement). A photon from a telescope strikes the photocathode which releases an electron. The dynodes are shaped and positioned so that the electrons are chanelled towards the next dynode. The photocathode is held at a negative potential of some 100 V with respect to earth. Each successive dynode is more positive than the photocathode by 100 V or so. The chain of dynodes produces an avalanche of electrons by secondary emission, which are collected to produce an output current proportional to the brightness of the object

A photomultiplier tube (PMT) consists of a series of **dynodes** arranged in an evacuated glass tube (Figure 3.12). Light falling on a photocathode ejects electrons by the photoelectric effect which are then accelerated through a potential difference to a dynode where more electrons are released by secondary emission. These are then accelerated to a second dynode and so on. As a result an electron amplification process occurs at each dynode, so that at the final collector electrode a useful current is produced. Photomultipliers do not produce images but measure *intensities* with a QE value as high as 20%. In general, the current produced by the photomultiplier is proportional to the number of detected photons. A telescope with a PMT attached at its focus enables weak stellar intensities to be measured.

Gamma ray detection

PMTs are used in conjunction with sodium iodide (NaI) crystals to detect gamma rays. When struck by a gamma ray, the sodium iodide crystal emits a flash of light which is detected by the PMT and amplified, signalling that a gamma ray event has occurred. The gamma ray detectors on the CGRO detect gamma ray bursts in this way.

Charge-coupled devices (CCDs)

A charge-coupled device (CCD) is a type of microchip in which light is converted directly into digital information. CCDs are made out of a silicon wafer divided into small regions called **pixels**. A typical CCD may have as many as a million pixels extending over an area of a few cm^2 arranged in rows and columns (Figure 3.13). When light strikes the CCD, electric charge

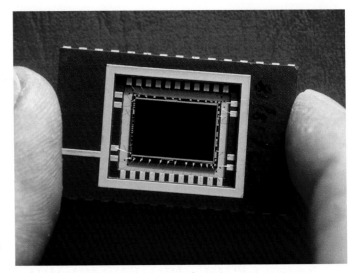

Figure 3.13 A CCD

is accumulated in the pixels which is proportional to the brightness of the image at a particular pixel location. Unlike photographic film, this makes the response of a CCD *linear*. The charge in each pixel is coupled to detecting electrodes and read out digitally for processing by a computer.

The quantum efficiency of a CCD is close to 100%, making it nearly an ideal radiation detector. In addition, it is able to detect a wide range of wavelengths from low energy X-rays to infrared. Its high QE means that the time needed to acquire an image of the same intensity relative to other image detectors is much smaller so CCDs require shorter exposure times, and that the light gathering power of a smaller telescope equipped with a CCD as a detector gives comparable performance to a much larger one using photographic film. These features, and the fact that CCDs output their information digitally, make them ideally suited to computer processing and data transmission. The images you see taken by the HST are recorded using CCDs. Its instruments, the WFPC and FOC, employ 800 × 800 pixel CCD cameras, and the readouts are transmitted to Earth in digital form ready for image processing.

Bolometers

A bolometer is a device that measures increases in temperature due to the radiant energy it receives at all wavelengths. There are two types: **thermistors**, which operate at room temperature, and **semiconductor bolometers** made of germanium, which need to be cooled to very low temperatures of the order of 2 K. For both kinds, as the device heats up, its electrical resistivity changes and so its resistance to an electric current will also change. The variation of current with temperature indicates how much energy the bolometer is absorbing. Bolometers are extensively used in infrared astronomy; the IRAS telescope used infrared semiconductor bolometric detectors which were cooled by liquid helium to 2 K, and operated for nine months before all the helium evaporated.

Analysing radiation from space

As well as detecting radiation from space, the astrophysicist needs to analyse it so that information about the physical processes occurring inside the object can be understood. There are four main types of analysers used for astrophysical measurements:

1 spectroscopes for measuring the spectra of starlight
2 photometers for measuring the brightness of stars
3 polarimeters for measuring the degree of polarisation of starlight
4 interferometers for high precision angular measurements of the separation of double stars and the diameter of single ones.

Spectroscopes

Spectrometers are instruments that split light into its component wavelengths. This is accomplished by using prisms or diffraction gratings to disperse light into a spectrum. Figure 3.14a shows the main features of a **prism spectroscope**. Light from a star passes through a slit where it is collimated before passing through the prism. The resulting spectrum is then brought to a focus on a detector, which is usually photographic film but can be a CCD. Figure 3.14b shows typical examples of stellar

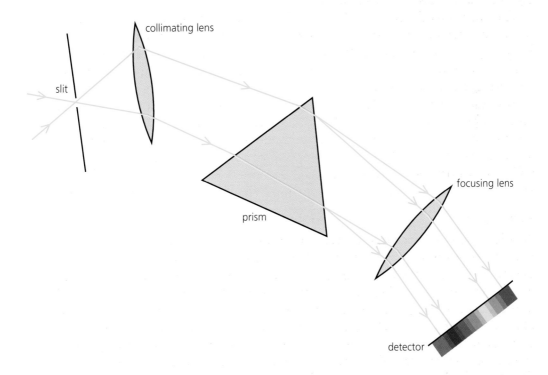

Figure 3.14 (a) A prism spectroscope

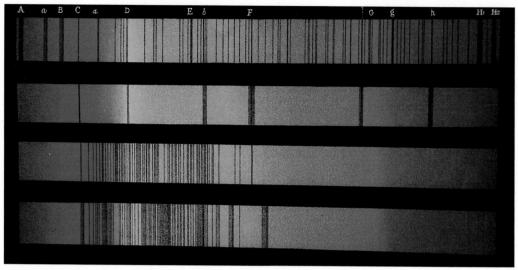

Figure 3.14 (b) Typical line spectra of the Sun, *Sirius*, *Aldebaran* and *Betelgeuse*

spectra. The appearance of vertical lines are due to the slit acting as a straight line source of starlight with characteristic wavelengths which is focused onto the image plane. This is why a spectrum like this is often called a **line spectrum**.

For light of a particular wavelength λ, the performance of a spectrometer is governed by two criteria:

dispersion = the change in wavelength over a distance Δx along the spectrum

$$= \frac{\Delta\lambda}{\Delta x}$$

and

spectral resolution $R = \dfrac{\text{the wavelength}}{\text{the smallest difference in wavelength that can be resolved}}$

$$= \frac{\lambda}{\Delta\lambda}$$

Higher dispersions increase the amount of detail that can be seen in the spectrum but also spread out the light, so high dispersion spectra can only be obtained for the brighter stars using telescopes with large apertures. Typical dispersions for astronomical work range from 200 nm mm^{-1} to 0.01 nm mm^{-1} and spectral resolutions range from 10 to 10 000. The detection element of a spectroscope usually has an electrical output that drives a graph plotter to produce a spectral profile on a chart recorder. In this case the analyser is referred to as a **spectrometer**.

Some of the disadvantages of prism spectrometers stem from the fact that relatively low levels of starlight pass through a thick prism and much of it is absorbed. In addition, the spectrum of light is not dispersed evenly – the red end is compressed while the blue and violet is spread out. For these reasons, another type of spectrometer commonly used in astronomical work is the **grating spectrometer**. This uses a diffraction grating as the dispersion element which consists of a glass slide onto which thousands of closely spaced parallel lines are cut, through which light is allowed to pass (a transmission grating). Alternatively, the grating can be ruled on a reflecting material from which the starlight is reflected (a reflection grating). In both cases the light is diffracted and different wavelengths constructively interfere along different directions. The resolving powers of gratings are generally better than those of prisms and, by using reflection gratings, it is possible to analyse wavelengths whose light would normally be absorbed by a glass prism.

Photometers

Measuring the light intensity or magnitude of stars with detectors is called **photometry** and an instrument which does this is called a **photometer**. The eye, photographic film, photomultipliers, CCDs and bolometers can all be used for photometric measurements which, in general, are based on the principle of *equalisation*, i.e. the observer has to judge when the star appears to be the same brightness as a reference star when light from either is passed through a filter or when their images are electrically calibrated using a scanning device.

Earlier in this chapter we mentioned the bolometric magnitude as the apparent magnitude of star measured above the Earth's atmosphere at all wavelengths (see page 56). However, in practice, astronomers measure the radiant energy of a star through three optical filters, called the *UBV* system, which are transparent to wavelengths centred on 365 nm (Ultraviolet filter), on 440 nm (Blue filter) and on 550 nm (Visual filter). A measurement of magnitude taken with the V filter would therefore roughly correspond to that estimated by the human eye.

In Chapter 2 we saw that Wien's Law tells us that the colour (i.e. dominant wavelength) of an object gives us information about its temperature (see page 30). Likewise, astronomers use photometry to determine the colour of a star so that they can estimate its surface temperature. The temperature of a star is thus determined by aiming a telescope at it and measuring the intensity of the light as it passes through each of the filters in turn. The relative intensities in neighbouring wavelengths are compared by subtracting the ultraviolet and blue magnitudes ($U - B$) and the blue and visual magnitudes ($B - V$). These are called the star's **colour indices** and, using the laws of blackbody radiation, it can be shown that a mathematical relationship exists (the derivation of which is beyond the scope of this book) between the surface temperature of a star and these indices.

Photographic methods

The images of stars on a photographic plate can be measured photometrically using an instrument called a **microdensitometer**. This consists of a device which shines a beam of light through the image and measures the transmitted intensity as a variation in voltage. The changing voltage is then displayed as a curve which records a continuous trace of intensity versus distance across the image. Another way of measuring magnitude is to see if there are stars on the plate of known magnitude, from which the unknown ones can be calibrated. It is also important to know fully the properties of the film emulsion such as the developing process, manufacture, age, wavelength sensitivity and temperature, as these will all be factors in making accurate measurements.

Polarimeters

Light from distant stars is polarised as it passes through clouds of interstellar gas and dust in the Milky Way, due to the presence of magnetic fields. The degree of polarisation is measured with a **polarimeter** and from this we can deduce the magnetic field strength existing inside the clouds. Polarimeters contain components which are **optically active**, that is, they alter the state of the incident polarised starlight. Polarisation elements usually consist of crystals such as calcite and special polarising materials which are used with focusing optics and filters to determine the degree of polarisation. The exact operation of these devices is technically complex, but it is sufficient to say that useful information about interstellar and galactic magnetic fields can be gained using polarimeters attached to a telescope.

Interferometers

We have already mentioned interferometers in connection with radio astronomy. There is another type of interferometer, the **Michelson stellar interferometer**, which is capable of measuring the angular separations of stars and even the angular size of the nearer stars. The stellar interferometer works in the same way as the Young's double-slit experiment. In Young's experiment, light is diffracted at an angle of θ as it passes through two slits separated by a distance d. If the path difference $d \sin \theta$ is equal to a whole number of wavelengths then the two beams will interfere constructively to produce a maximum.

In a stellar interferometer the light source can be two stars, or light from opposite ends of a star along its diameter. The light from these sources is incident on two mirrors separated by a distance d and is brought to focus with a telescope. The sources produce *independent* fringe patterns of the same brightness that can be made to overlap and disappear by adjusting the distance between the mirrors. However, unlike Young's slits, the sources of light are *not* coherent. The reason the fringes vanish is *not* an interference effect but simply due to the pattern of light and dark fringes from the two sources overlapping each other.

If we are looking at a single star with an angular size (in radians) of α as a two-dimensional circular disc, then the interference pattern vanishes when $\alpha = 1.22 \; \lambda/d$.

WORKED EXAMPLE 3.8

Using a stellar interferometer, a star is found to have an angular size of $\alpha = 28 \times 10^{-3}$ ". If it lies at a distance of 22 pc, what is its radius?

First of all, we have to convert arcsec into radians and parsec into m:

$$28 \times 10^{-3}\ " = 1.4 \times 10^{-7}\ \text{rad}$$
$$22\ \text{pc} = 6.8 \times 10^{17}\ \text{m}$$

Using the fact that length of an arc = arc radius $\times$ subtended angle (radius)

$$\tan \alpha \approx \alpha = \frac{\text{diameter of star}}{\text{distance of star}}$$

then the star's diameter is:

$$(6.8 \times 10^{17}\ \text{m}) \times (1.4 \times 10^{-7}\ \text{rad}) = 9.5 \times 10^{7}\ \text{km}$$

and its radius is 4.8×10^{7} km.

Image processing

In the last 20 years there has been a revolution in computing power, with computers increasing in processing ability while at the same time decreasing in cost. This, together with technological advances in acquiring astronomical data in digital form using detectors such as CCDs, means that the manipulation of image data inside a computer has now become a routine tool in extracting the maximum amount of information from an astronomical image. **Image processing** is used for removing known imperfections and enhancing the visual appearance of the object as presented to the human eye. Some standard image processing techniques are described in this section.

Noise removal

Bright spots may appear in the image due to cosmic rays striking the detector or faulty pixel elements in a CCD. These sources of **noise** may be mistaken for real features but can be digitally removed and replaced with the average intensity of their adjacent areas.

Background subtraction

Even when a detector is not exposed to an image there will always be some residual exposure, either due to background illumination or, in the case of a CCD, a 'dark current' due to electrons accumulating in pixels as a result of thermal energy (CCDs are cooled to reduce this effect). If the extent of this background signal is known, then it is a systematic error that can be subtracted out by the computer in the final image.

Contrast stretch

Consider a computer that can display 16 levels of grey while a typical CCD can resolve 256 changes in intensity. In order to make use of the CCD's greater range of levels, a **contrast stretch** is applied. This involves mapping the dominant grey levels in the image to the maximum range of the computer display.

To see how this works, suppose 80% of pixels in a CCD have grey levels between 140 and 172. This represents a range of 32 levels in which most of the useful information is contained. If we then map CCD level 140 and 141 to level 0 of the computer display, levels 142 and 143 to level 1, and levels 144 and 145 to level 2 and so on (Figure 3.15), then the computer's display range is used in the most efficient manner. Modern computers can easily display 256 levels of grey, but the same principle described in the simplified diagram in Figure 3.15 still applies, enabling an even greater range of detail to be imaged using state-of-the-art computer workstations and better resolution CCDs.

Figure 3.16 shows the results of a contrast stretch applied to an image of the spiral galaxy M96. Notice how more detail of the spiral arms is revealed in (b) as compared with (a).

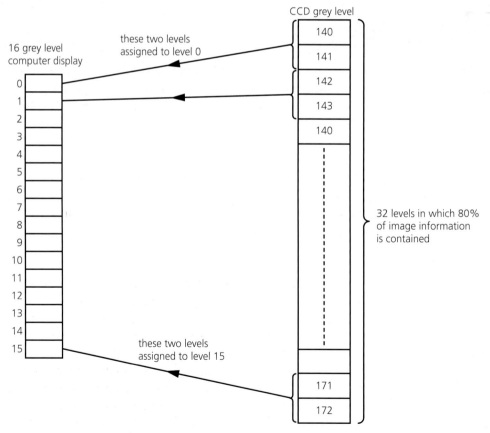

Figure 3.15 A contrast stretch

(a) Original image   **(b)** Image after applying a contrast stretch

Figure 3.16 The effect of a contrast stretch on the image of the spiral galaxy M96 in the constellation *Leo* (image obtained using the 2.5 m Isaac Newton Telescope on La Palma).Courtesy of Nial Tanvir Institute of Astronomy, Cambridge

False colour

Sometimes it is helpful to assign colours from the computer's colour palette to particular intensities on the image. This is particularly useful when astronomers want to find, for example, the temperature distribution of an object or the intensity variation of radiation across the sky.

Other techniques

Astronomers use many other image processing techniques in addition to those described above. Some involve enhancing the edges of individual features or correcting for distortions introduced by the detector and observing instrument, while others, such as filtering and three-dimensional projection, involve advanced mathematical methods and are not discussed further in this book. Processing techniques may be applied individually or sequentially (for example a contrast stretch followed by a false colour assignment), and separate images may be combined in various ways. However, human judgement is always needed to determine which technique is appropriate for a particular object in order that the information is interpreted correctly.

Summary

◆ Astrophysicists use special units to measure astronomical quantities. These include the **astronomical unit**, the **light year** and the **parsec** for distances, the **Jansky** for radio power per unit area per Hz, and units of radius, mass and luminosity expressed as multiples of the Sun's values. The parsec is defined by the method of **parallax** which involves using trigonometry to calculate the distances to the nearer stars.

◆ The brightnesses of stars are measured using a system of **magnitudes** based on a number scale. The more negative the number, the brighter the star appears. The **apparent magnitude** is a measure of a star's *observed* brightness. The **absolute magnitude** is the apparent magnitude of a star if located at a distance of 10 parsecs from the Earth.

◆ **Pogson's Law** relates apparent magnitude m to absolute magnitude M, and is written as $m - M = 5 \log d - 5$ or, alternatively, $d = 10^{(m-M+5)/5}$ where d is the distance in parsecs. The relationship between a star's luminosity and its absolute magnitude is given by $M = 4.77 - 2.5 \log(L/L_\odot)$.

◆ The **bolometric magnitude** is a star's apparent magnitude measured above the Earth's atmosphere over all wavelengths.

◆ The **period–luminosity (P–L) relation** is a graph that shows how the luminosities of Cepheid variable stars vary with time. By measuring the period, the absolute magnitude of a Cepheid can be determined and hence its distance using Pogson's Law. Cepheids are important distance indicators for measuring astronomical distances.

◆ Most of the information that reaches us from astronomical objects comes to us in the form of electromagnetic radiation, and the atmosphere of the Earth limits the range of wavelengths that can be collected from the ground. Ground-based collectors include optical and radio telescopes of different designs and focusing techniques. **Space astronomy** is the use of balloons, rockets, satellites and spacecraft which carry instruments that can observe space at wavelengths normally absorbed by the Earth's atmosphere.

◆ Various kinds of radiation detectors are employed in astronomy. These include photographic film, thermistors, CCDs, particle detectors and light amplification devices. An important measure of a detector's usefulness is its **quantum efficiency**.

◆ Radiation analysers are instruments that are able to measure and separate radiation according to its form and wavelength and include spectrometers, photometers, polarimeters and interferometers. **Image processing** is the manipulation of images using a computer in order to extract the maximum information from an astronomical image. Important image processing techniques are noise removal, background subtraction, contrast stretching and false colour assignment.

Questions

1 a Table 1 gives values of the masses of objects in the universe. The masses are quoted in terms of $M_\odot$, the mass of the Sun.

Table 1

Mass/$M_\odot$	Identity
1×10^{-10}	
1×10^{-5}	
1	The Sun
20	

Complete Table 1 by suggesting what the objects might be.

b Table 2 gives distances from Earth, or diameters, of objects in the universe.

Table 2

Distance (or diameter)	Identity
1.5 light seconds	
8 light minutes	Distance between the Earth and the Sun
6×10^4 light years	
1×10^7 light years	

Complete Table 2 by suggesting what the distances or the diameters might be.

UCLES, June 1998

2 The luminosity of the Sun is 3.90×10^{26} W. Assuming we regard the Sun as a point source of light, how much dimmer would the Sun appear to an observer on Pluto at a mean distance of 39.44 AU compared with its appearance to an observer on Earth?

3 a The distance of a star from the Earth may be found directly by the annual parallax method or indirectly from intensity measurements when the absolute luminosity (magnitude) is known.

Explain briefly the principle of the annual parallax method.

How might a value for the absolute luminosity (magnitude) of a star be calculated without its distance being known?

Explain how stellar distance can be found from intensity measurements.

Explain why you need to use both methods of measuring stellar distance when exploring the structure of our local galaxy.

ULEAC, June 1998, part

b The star *Vega* has an annual parallax of 0.133 arcseconds. What is its distance in

i) parsecs   ii) AU   iii) light years?

4 a What is the difference between apparent and absolute magnitude?
The more negative the magnitude of a star is, then the brighter it is. The Sun has an apparent magnitude of -26.74 at 1 AU. Why therefore, does it have a *positive* absolute magnitude of $+4.77$?

b The brightest star in the night sky is *Sirius* with an apparent magnitude of -1.5. Calculate the distance of the Sun at which its apparent magnitude would be equal to that of *Sirius*.

5 a The star *Altair* has an apparent (visual) magnitude of $+0.77$. If *Altair* lies at a distance of 16 ly, what is its absolute magnitude?

b The star *Regulus* has an apparent (visual) magnitude of $+1.35$ and an absolute magnitude of -0.3. How far away from us is it in parsecs?

c The star *Capella* is approximately 100 times more luminous than the Sun. What is its absolute magnitude? If its apparent (visual) magnitude is measured as $+0.08$, how far away is it?

6 a The diameter of the human eye pupil in the dark is about 0.5 cm. A reflecting telescope has an objective mirror of diameter 20 cm and focal length of 240 cm. What is its light gathering power (LGP) relative to the human eye and its magnification when used with eyepieces of focal length
i) 6 mm
ii) 8 mm
iii) 24 mm?

b The Hubble Space Telescope (HST) has a circular objective mirror of 2.4 m. What is its light gathering power? Calculate the theoretical image resolution of the HST at $\lambda = 500$ nm (visible), $\lambda = 200$ nm (ultraviolet) and $\lambda = 2000$ nm (infrared).

7 Name the two main wavelength bands of the electromagnetic spectrum which are used for terrestrial astronomy. What property of these wavebands makes them particularly suitable for this purpose?
State and explain *two* different benefits of observing from above the Earth's atmosphere.

ULEAC, Jan 1999, part

8 The transparency of the Earth's atmosphere affects the locations at which astronomical observations in different regions of the electromagnetic spectrum are possible.
Write 'yes' in the relevant places below to indicate at which locations observations may be made.

	γ-rays	X-rays	visible	microwave	radio
At sea level					
Aboard a space probe					

UCLES, June 1997

9 What is *image processing*? Explain two image processing techniques and how they might be used to enhance astronomical images.

10 a What is the main physical reason that limits the image resolution of a radio telescope?

b A very large baseline interferometer whose antennae are separated by 30 km is observing an object at a wavelength of 1.5 cm. What is the theoretical image resolution? Suppose the antenna were placed at opposite points on the equator. What then would the theoretical resolving power at this wavelength then be? (radius of Earth = 6400 km)

11 a The distance of *Sirius A* from the Sun, calculated from the measured value of its annual parallax, is 8.7 light years. One light year equals 9.46×10^{15} m.

Explain the term *annual parallax* by sketching and labelling a suitable diagram.

How would the annual parallax of *Sirius A* be measured if the distance of the Earth from the Sun is already known? On what assumption does the method depend?

The distance of the Earth from the Sun is 1.50×10^{11} m. Calculate the annual parallax for *Sirius A*.

b Explain what is meant by the *luminosity* of a star.

The luminosity of *Sirius A* is 8.17×10^{27} W. Calculate the *intensity* of *Sirius A* measured at the Earth.

London, June 1999, part

Stellar spectroscopy

Astrophysicists can learn a great deal about the nature of a star by analysing the light received from it. Information such as its temperature, physical nature and chemical composition can be deduced from the pattern of spectral lines that are evident in the star's spectrum. The study of stellar spectroscopy marked the first real beginnings of astrophysics.

The message of starlight

More than 300 years ago Sir Isaac Newton showed that sunlight can be split into different colours using a prism. He found that the shorter the wavelength the greater the angle of refraction, so that blue light, for example, is refracted more than red. The effect of this is to form a **spectrum** of light from red through to violet. Stellar spectroscopy is the study of the spectra of starlight. It is a very powerful tool that enables astrophysicists to infer many physical and chemical properties of stars and classify them on the basis of their spectral characteristics.

Stellar spectra are analysed using spectroscopes. In Chapter 3 we saw that their operation is based on optical principles, either of refraction as in the case of a prism, or of interference in the case of a grating. In order to understand how spectroscopy can be a useful tool to astrophysicists we need to describe the different kinds of spectra that are observed and explain how they arise.

Kinds of spectra

(a) Continuous spectrum

(b) Bright line or emission spectrum

(c) Absorption spectrum

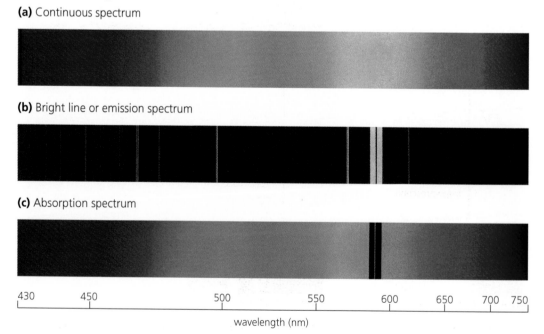

wavelength (nm)

Figure 4.1 The general appearance of the three kinds of spectra

The German physicist Gustav Kirchhoff studied the properties of spectra in the laboratory and discovered that there are three kinds (Figure 4.1). The different kinds are produced under different physical conditions inside the source of light. From these investigations, Kirchhoff formulated three empirical rules of spectral analysis:

Rule 1   A hot opaque solid, liquid or gas which is under *high pressure* will emit a **continuous spectrum**.

Rule 2   A hot gas *under low pressure* (i.e. much less than atmospheric) will emit a series of bright lines on a dark background. Such a spectrum is called a **bright line** or **emission spectrum**. The number and position of the bright lines depends on the chemical composition of the source.

Rule 3   When light from a source that has a continuous spectrum is shone through a gas at a *lower temperature and pressure* then the continuous spectrum will be observed to have a series of dark lines superimposed on it. This kind of spectrum is known as a **dark line** or **absorption spectrum**. The number and positions of these dark lines depend on the chemical composition of the *cooler* gas.

These rules are illustrated in Figure 4.2.

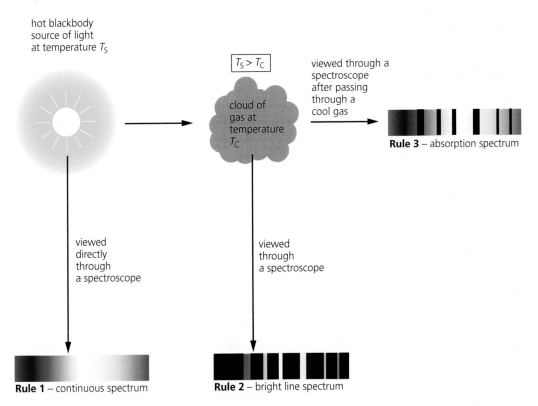

Figure 4.2 Kirchhoff's three rules of spectral analysis. Note that the dark lines in the absorption spectrum are at the same wavelengths as those in the bright line spectrum

Atomic processes

The Bohr model of the atom can be used to understand how these three types of spectra are produced. In Chapter 2 (page 26) we saw that photons are absorbed or emitted when electrons are excited to higher energy levels or drop down to lower ones. The spectral lines of the hydrogen atom were described and the important fact that emission or absorption of light *only* takes place when a photon has an energy equal to the energy difference between quantised energy states of an electron in the atom.

Continuous spectra

In a high pressure hot gas, the atoms have high kinetic energies and collisions between them are very frequent. Their electrons are raised to high energy excited states and then relax to lower energy levels, their energy emerging as discrete photons. However, if the gas is at very high pressure and density, then an electron in its excited state may not have enough time to drop down to its ground state before the atom undergoes another collision with a neighbouring atom. This has the effect of blurring the sharpness of each emission line into a broad band of wavelengths. The same effect happens to all other lines so that by the time the light emerges from the gas it has 'smeared out' into a continuous spectrum at all wavelengths (see Figure 4.1a).

Emission spectra

In a gas containing atoms of only one kind, all the electrons will be in their ground state if the temperature of the gas is very low. As the gas is heated, its atoms gain kinetic energy and collide with their neighbours, causing their electrons to be raised to excited states. As the electrons drop down, photons are emitted of many different energies and wavelengths according to the particular electron energy level pattern for the gas.

The emission of these lines will cause the gas to glow with a light composed of wavelengths that correspond to the electron energy transitions. For moderate temperatures we might find that the electrons only reach the first excited state of the atom, in which case the emitted light when observed through a spectroscope will consist of a single bright emission line corresponding to the difference in energy between the first excited and the ground state.

As the temperature is increased, more emission lines will start to appear until at higher temperatures many lines will be visible, corresponding to all the allowed energy transitions of electrons in the gas. In this way an emission or line spectrum is formed that is related to the elemental composition of the gas (see Figure 4.1b).

Absorption spectra

To explain Kirchhoff's third rule, we need to consider what happens when we place a gas of unknown composition in front of a source of light that emits a continuous spectrum. As we have seen, light from a continuous source contains photons of all energies and wavelengths. Now if it is the case that the energy of some of these photons is exactly equal to the difference between the ground state and an excited state of an atom in the unknown gas, then that photon will be absorbed by an electron in the unknown gas, which is then excited to a higher level.

The excited electron will quickly return to the ground state, emitting a photon. The emitted photon need not be emitted along the same direction as the one that was absorbed; it is usually emitted in a *different direction*. The re-emitted photons are therefore not generally detected when looking through a spectroscope at the source, and the continuous spectrum is observed to have dark lines at the wavelengths corresponding to transitions between the energy states of atoms in the unknown gas (see Figure 4.1c). It follows that *it is precisely these same wavelengths which would be emitted in an emission spectrum if the unknown gas were heated to a high temperature.*

This is a very important finding. If we photograph the continuous spectrum marked by dark lines and place it next to a photograph of the emission spectrum of the unknown gas then we find that every dark line present in the continuous spectrum will match a bright line in the emission spectrum (Figure 4.3). Therefore, we have a way of determining the elemental composition of a gas from its spectrum. Both the dark lines superimposed on the continuous spectrum and the bright lines in the emission spectrum provide a 'spectral fingerprint' that identifies the elements present in a gas.

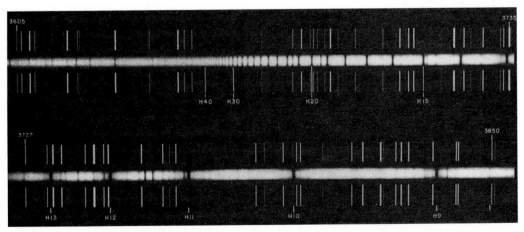

Figure 4.3 The Balmer series of spectral lines as seen in the spectrum of a star. It can be clearly seen that for the hydrogen lines H15, H20 and H30, bright line comparison spectra appear above and below the stellar spectrum. Other bright lines corresponding to hydrogen absorption lines in this photograph are too faint to be seen

The spectrum of the Sun

The first star to be studied spectroscopically was the Sun. A British astronomer William Hyde Wollaston (1766–1828), using a prism, observed that the Sun emitted a continuous spectrum that had a series of dark lines superimposed on it. Using a diffraction grating a German telescope maker Joseph von Fraunhofer (1787–1826) repeated Wollaston's observations and found 784 dark lines which are now known as **Fraunhofer lines**. Fraunhofer realised that some of these dark lines had the same wavelength as bright emission lines of spectra of various elements which were studied in the laboratory. An example was iron, which when heated to a gaseous state displays bright emission lines that can be matched to dark ones in the solar spectrum.

Astrophysicists have now observed thousands of dark absorption lines in the Sun's spectrum. Using Kirchhoff's rules they have been able to detect the presence of some 67 different elements in the Sun. Figure 4.4 shows part of the Sun's spectrum. Several hundred lines are visible and some corresponding to various elements are indicated.

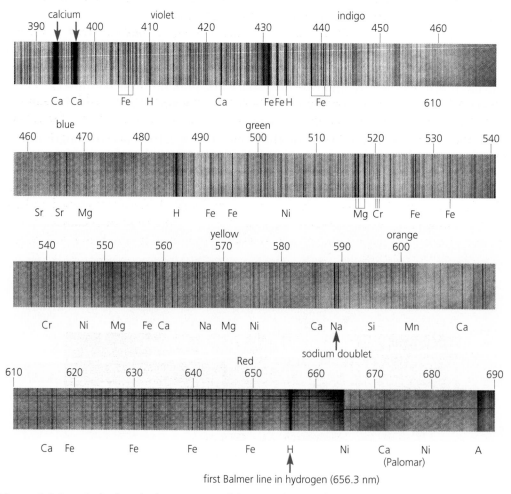

Figure 4.4 Fraunhofer lines in the spectrum of the Sun. The wavelengths indicated are in nm

As an interesting footnote, the element helium was first discovered in the Sun before it was found on Earth. An unknown line in the Sun's spectrum was observed which could not be related to lines of elements then known in the laboratory. It was named after the Greek word for the Sun, *helios*, and was subsequently discovered on Earth 40 years later!

The spectra of other stars

When we observe the spectra of other stars we find that some are similar in appearance to the Sun's but others are very different. For example, Figure 4.5 shows the absorption spectrum of the star *Vega* compared with that of the Sun in the blue–violet region. *Vega* is a very hot star in the constellation *Lyre* and a pair of binoculars will easily show it glowing with a bluish tinge. The Sun's spectrum shows two lines of hydrogen at 410.1 nm and 434.0 nm plus many other spectral lines. In the spectrum of *Vega* these same two lines are much wider and more intense. *Vega*'s spectrum also has a marked absence of other dark lines compared to the Sun.

At first sight you might think that the thicker hydrogen lines mean that there is a greater abundance of hydrogen in *Vega* than in the Sun. Actually, this is *not* the case and it is a remarkable fact that the compositions of most stars are broadly similar in their chemical mixtures. The reason is that the width of the lines is related to the star's *temperature*. The broadening of a line as the temperature is raised is a result of higher collision rates blurring the transition energy, and the wider range of speeds of emitting atoms causing a range of Doppler shifts in the wavelength of the emitted light (see pages 94–5).

The temperature also controls the relative intensities of the various lines in the spectra of different stars. To see why, consider the two hydrogen lines in Figure 4.5. The 410.1 nm line is produced when a photon of energy 3.02 eV is absorbed and an electron jumps from energy level $n = 2$ to $n = 4$; the 434.0 nm line is caused by a photon of energy 2.85 eV being absorbed when an electron jumps from $n = 2$ to $n = 5$ (see Figure 2.3, page 27).

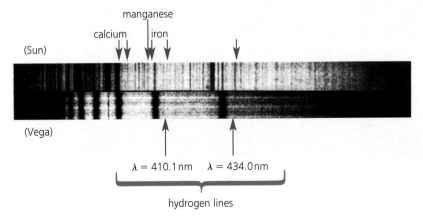

Figure 4.5 The absorption spectrum of the Sun (top) compared with that of the star *Vega* (bottom)

For these lines to be formed, the hydrogen atom must initially have had enough energy to be in its first excited state, about 10 eV above the ground state, and this energy will have come from the thermal motion of the atoms in the gas. However, a thermal energy of 1 eV corresponds to a temperature of about 10 000 K, so temperatures must be very high to produce appreciable excitation by heating alone. For cool stars whose surface temperatures are low to moderate, most of the hydrogen atoms will be in the ground state. As a result the Lyman series, which consists only of lines involving transitions from the ground state, will figure prominently in the spectrum. Lines corresponding to the Balmer series, such as the two in Figure 4.5, will be weak since the hydrogen atoms will have fewer electrons populating the $n = 2$ state. At higher surface temperatures, however, there will be a large number of electrons populating the first excited state and the frequency of transitions from $n = 2$ to higher states will be greater. As a result, these stars will show the 410.1 nm and 434.0 nm lines more strongly. Figure 4.6a shows the effect on these lines as the surface temperature of the star increases. However, look at Figure 4.6b. Above a temperature of about 15 000 K, the strength of the lines starts to *decrease* until for hot stars (above 35 000 K) they are comparatively faint. In the spectra of very hot stars (above 50 000 K) the lines are hardly visible at all.

The reason for this is that at extremely high temperatures the kinetic energies of the hydrogen atoms are so high that violent collisions take place causing them to become ionised. Stripped of their electrons, they are now simply protons and are unable to absorb or emit light. The line spectrum is therefore weak because only a small fraction of the hydrogen atoms in very hot stars are able to keep their electrons and remain in a non-ionised state. We say that the stellar material is now in a **plasma state**.

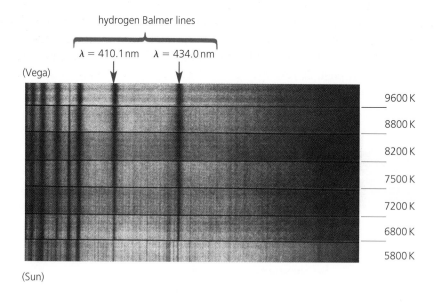

Figure 4.6 (a) As the temperature increases the intensities of individual lines increase

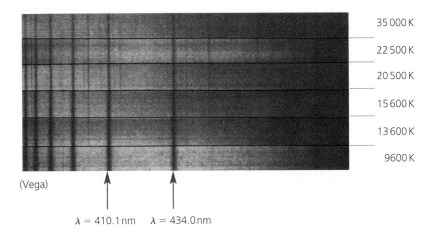

Figure 4.6 (b) At very high temperatures the intensity of the lines becomes weaker due to increasing ionisation of the hydrogen atoms

Molecular spectra

Some stars display spectral features which consist of many finely spaced lines in a band of wavelengths, caused by molecules in the outer layers of a star. While a molecule consists of atoms with excited electronic states, it also has collective rotational and vibrational motion which is also quantised. If an electron undergoes a transition within an atom in a molecule to an excited energy state then the rotational and vibrational energies also change although by a smaller amount. This has the effect of separating the energy of the electron's transition into many closely spaced energies and corresponding wavelengths (Figure 4.7).

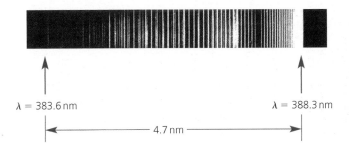

Figure 4.7 Molecular band spectrum

The shape of a spectral line

The appearance of a spectral line in a star's spectrum is influenced by a number of physical processes occurring in the stellar atmosphere. A line's shape or **profile** is described as being weak, strong, or very strong according to its intensity. The three main processes that effect the shape of a spectral line are called **collisional broadening**, **Doppler broadening** and **rotational broadening**. In addition, a lesser effect called the **Zeeman effect** can also cause splitting of the spectral lines and involves the star's magnetic field.

Collisional broadening

If two atoms collide, then the electrons on each atom will repel each other and distort their respective energy levels. If a collision happens when one of the electrons is absorbing a photon, then the absorbed photon energy will be altered from the value it would have had if the atom was left undisturbed. In a gas that is at a moderate temperature and density, collisions between atoms are infrequent and so absorption of a photon is likely to happen when the atom is undisturbed. At higher temperatures and pressures, collisions become more frequent and the photon energy is more likely to be affected. In this situation, the absorbed photon energies vary over a considerable range and therefore wavelength, and the spectral line is widened or **collision broadened** (Figure 4.8).

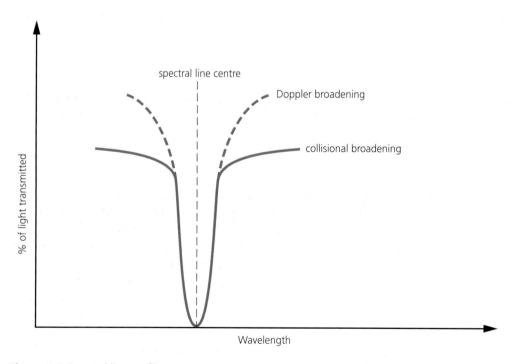

Figure 4.8 Spectral line profile

Doppler broadening

In the atmosphere of a star, the atoms have random velocities due to their thermal energy. At any instant some of the atoms travel towards us and others away when they emit photons. This produces a Doppler shift in the absorption lines of the spectrum. The effect on the shape of the line is different from that due to collisions. The Doppler broadening falls off more sharply from the line centre than does collisional broadening (Figure 4.8).

Rotational broadening

A star that is rotating will produce a Doppler shift in each line of the star's spectrum. Unless the axis of rotation is directly in the line of sight, then the atoms in the surface layers of the star will alternately move towards and then away from the observer. The amount of broadening depends on the rotation rate and on the angle of inclination of the axis of rotation relative to the line of sight. Using a **microdensitometer** together with mathematical models of line shapes, astrophysicists can deduce the line profile and calculate the stellar rotation rate. For simplicity, let's assume that the axis of rotation is perpendicular to the line of sight of the observer. If the change in wavelength of a line at wavelength λ is $\Delta\lambda$, then the velocity v of atoms on the limb of a rotating star is given by:

$$v = c \, \frac{\Delta\lambda}{\lambda}$$

If we know the radius R of the star then the period T of rotation can be calculated from:

$$T = \frac{2\pi R}{v}$$

Astrophysicists have found that, in general, the hottest stars rotate the fastest with periods as short as 4 hours. Stars like the Sun rotate fairly slowly, about once every 27 days.

The Zeeman effect

Electrons in atoms are moving charges that constitute 'rings' of electric current. This has an electromagnetic effect, producing a magnetic field similar to that of a bar magnet. When these tiny atomic magnets are placed in a strong external magnetic field, the electron shell structure of the atom is distorted and the energy levels become shifted. This causes the spectral lines to shift and this **Zeeman effect**, named after the Dutch physicist Pieter Zeeman (1865–1943), may be directly observed in the spectral line pattern. It is possible to relate the degree of 'splitting' (or changes in wavelength) to the strength of the external magnetic field and, in this way, astrophysicists can obtain information about a star's magnetic field strength. Zeeman splitting is particularly useful in the study of sunspots which have very intense magnetic fields and produce pronounced splitting in the absorption spectrum of the Sun.

Classifying stellar spectra

The changes in intensity of the hydrogen lines with temperature enable us to devise a spectral classification system. The first person to attempt to do this was the Italian astronomer P.A. Secchi, who in 1860 classified stars into four distinct groups based on their spectral features. The modern version is called the **MK system** (devised by and named after W.W. Morgan and P.C. Keenan). To be classified, a star is assigned a **spectral type** and a **luminosity class**.

Spectral types

The spectral type of a star is designated by one of seven letters O, B, A, F, G, K, M, ranging from the hottest type (O-type) to the coolest type (M-type). Table 4.1 shows the temperatures and characteristic features in the star's spectrum that distinguish the spectral types.

Table 4.1 The MK spectral classification system

Type	Surface temperature/K	Spectral feature
O	> 20 000	Ionised helium (He II)
B	10 000–20 000	Neutral helium, hydrogen lines start to appear
A	7000–10 000	Strong neutral hydrogen (Balmer lines) visible
F	6000–7000	Ionised calcium (Ca II) visible, hydrogen lines weaker
G	5000–6000	Ionised Ca II very prominent, much weaker neutral H lines, also other metallic lines such as iron
K	3500–5000	Neutral metals such as calcium and iron prominent, molecular bands visible
M	2000–3500	Molecular bands very visible, particularly those of titanium oxide (TiO)

Our Sun is a G-type star.

In practice the classification of stars into spectral types is more complex than this. Each type can be subdivided into at least ten subdivisions, so that one might refer to a star of type A5, lying halfway between type A0 and F0. The spectral type order is commonly remembered by the mnemonic 'Oh Be a Fine Girl (Guy) Kiss Me'!

Luminosity classes

For a given temperature, some stars are more luminous than others. This is usually because the star is larger and its outer atmosphere more tenuous and at lower pressure than a fainter star. The spectral lines of very luminous stars are much narrower since the effects of line broadening due to collisions is much less and the line profile is sharper. We can therefore further classify stars *for each spectral type* in terms of luminosity on the basis of 'sharpness' of their spectral lines. These **luminosity classes** are divided into seven principal star-types, denoted by roman numerals:

I   supergiant stars
II   bright giant stars
III   giant stars
IV   subgiant stars
V   main sequence dwarf stars
VI   sub dwarf stars
VII   white dwarf stars

In practice some luminosity classes, particularly those of the supergiants, are subdivided into suffixes *a*, *ab* and *b*, and a class written as III–IV means a star with characteristics midway between the two classes.

The full spectral classification thus consists of [Spectral type][number] [Luminosity class][suffix (if any)]. For example, the Sun is classified as a G2V star, *Canopus* (a supergiant) as A9II, and *Betelgeuse* (a red supergiant) as M2I*ab*.

The chemical composition of stars

It is clear that the spectral lines observed in a star's spectrum arise from the chemical elements present in the stellar material. Each element leaves its 'signature' in the form of a pattern of spectral lines corresponding to its electron shell structure. Strictly speaking, the spectral lines we observe correspond to material in the star's *atmosphere* and not its interior. However, for most stars, the hot interior gases convect upwards so that the outer layers become thoroughly mixed. In addition, the assumption is made that all stars form out of a homogenous cloud of materials and, as a result, the chemical composition of the stellar atmosphere is characteristic of the star as a whole.

Measurement of the relative abundances of the chemical elements in a star is complicated by the star's temperature. You may think that the more intense the spectral line pattern is, then the more of that particular element the star contains. However, we have seen that a faint set of absorption lines can be due to the fact that the temperature of a star is such that not all the electrons of a particular element are in the correct initial energy levels in order to produce a particular line. In the case of the Sun's spectrum, the 656.3 nm hydrogen line, produced when an electron in the first excited state ($n = 2$) is raised to the second excited state ($n = 3$), is faint. This is because at temperatures on the solar surface most electrons are in the ground state and only a small fraction are in the first excited state, and is *not* due to any lack of hydrogen!

In order to calculate the relative abundances of the chemical elements in the stars, the astrophysicist needs to know, for a particular element, what fraction of atoms are in the first excited state, what fraction in the second and so on. It is found that for the majority of stars the chemical composition is very nearly the same. By mass, most stars contain about 72% hydrogen, 25% helium and the remaining 3% is made up of other elements (notably iron) in roughly equal abundances.

Summary

◆ Stellar spectroscopy is the study of the spectra of starlight. There are three kinds of observed spectrum: **continuous, emission** and **absorption spectra**, which are produced under different physical conditions inside a star's atmosphere.

◆ **Kirchhoff's three rules of spectral analysis** tells us that:

1 A hot glowing gas under high pressure emits a **continuous spectrum** at all wavelengths.

2 A hot gas under low pressure will emit a series of bright **emission lines** on an otherwise dark background.

3 When a continuous spectrum is observed through a cooler gas, dark **absorption lines** appear in the continuous spectrum.

The bright lines in the emission spectrum of the gas occur at the same wavelength as the dark lines in the absorption spectrum of the same gas. For a hot glowing object surrounded by 'background' gas, absorption lines are seen in the spectrum if the background gas is cooler. If hot, the background gas will produce emission lines at the same wavelengths.

◆ By comparing spectral line patterns observed in a laboratory with those from starlight, Kirchhoff's three rules can be used to determine the elemental composition of stars. The chemical compositions of most stars are found to be broadly similar.

◆ The intensity or strength of a spectral line in a star's spectrum is related to the star's temperature. Some spectral lines are split up into a band of discrete wavelengths due to molecular energy levels. Others are split by the **Zeeman effect** due to the presence of magnetic fields.

◆ The shape of a spectral line is called its **line profile**. Line profiles may be broadened by the processes of **collisional, Doppler** and **rotational broadening** and from these astrophysicists can deduce information about gas pressures, particle velocities, and the rotation of a star.

◆ The changes in intensity of the Balmer lines with temperature enable astrophysicists to classify stars by **spectral type**. In addition, for the same spectral type, stars can be further divided into seven **luminosity classes** based on the sharpness and relative widths of their lines.

Questions

1 Distinguish between continuous, emission and absorption spectra.

2 The spectrum of sunlight has dark lines in it. One of these lines occurs at a wavelength of 588 nm. Calculate the energy of a photon with this wavelength.
   The diagram below shows the energy levels of the helium atom.
   Explain with reference to these energy levels how the dark line in the spectrum may be due to the presence of helium in the atmosphere of the Sun.

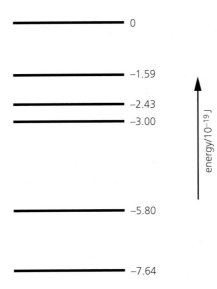

3 The Balmer series of spectral lines are emitted as electrons fall to the $n = 2$ orbit. It is found that the wavelength λ of the Balmer lines can be expressed by a remarkably simple formula:

$$\frac{1}{\lambda} = R\left(\frac{1}{2^2} - \frac{1}{n^2}\right) \qquad n = 3,4,5.....$$

where $R = 1.0974 \times 10^7 \text{m}^{-1}$ and is called the **Rydberg constant** after the Swedish spectroscopist Johannes Robert Rydberg (1854–1919) who determined its value.
   When n is set from three to infinity then this formula yields the Balmer series of spectral lines. Calculate:

a the first ten lines of the Balmer series

b when $n = \infty$, what is the limit of wavelength of the Balmer series?

4 a Why do the observed spectral lines of stars become weaker the higher the temperature of the star?

b What three processes affect the shape of a spectral line?

5 Classify the following stars in ascending order of temperature. Which one is most like the Sun?

Star	Spectral type
Arcturus	K2III
Spica	B1V
Antares	M1Ib
Alpha Centauri A	G2V
Sirius	A1V
Fomalhaut	A3V

Observational properties of stars

Stars have a range of observed properties by which they are characterised. There are two significant relationships that emerge from these properties, connecting a star's luminosity with its temperature and with its mass. By observing stars, an important diagram called the Hertzsprung–Russell Diagram can be derived which demonstrates that there are different kinds of stars and that they can be put into groups based on their physical features.

Six observational properties

The main observational properties that characterise a star are its:

1 luminosity   2 temperature   3 radius   4 mass   5 chemical composition   6 age

In this chapter we will see how these properties are determined observationally and how it is possible to group stars together on the basis of their similarities. Table 5.1 shows these properties listed for the Sun.

Table 5.1 Six observational properties that characterise the Sun

Luminosity	3.83×10^{26} W	Mass	1.99×10^{30} kg
Temperature (photosphere)	5780 K	Chemical composition	Mainly hydrogen and helium
Radius	6.96×10^{8} m	Age	About 5×10^{9} years

Luminosity

The luminosity of stars can be measured from their apparent magnitude using Pogson's Law if their distance is known and also, as we saw in Chapter 4, their spectra enables us to assign them to a luminosity class. The values of stellar luminosities range from $10^{-4} L_{\odot}$ up to $10^{6} L_{\odot}$ and Table 5.2 shows luminosity values for six stars within 1000 light years of the Sun.

Table 5.2 The range of stellar luminosities

Star	Luminosity/$L_{\odot}$	Distance/ly
Antares	3×10^{6}	420
Rigel	10^{5}	880
Arcturus	100	36
Altair	10	16
Alpha Centauri	2	4
Barnard's Star	10^{-2}	6

The Sun falls about halfway between the brightest and faintest stars; however, its power output per unit mass is relatively small. The luminosity of the Sun per unit mass is only

$$\frac{3.83 \times 10^{26} \text{ W}}{1.99 \times 10^{30} \text{ kg}} \approx 2 \times 10^{-4} \text{ W kg}^{-1}$$

Compare this with the average power output per unit mass for a human being, which is 1 W kg^{-1}! Clearly you and I are far more powerful per unit mass than the Sun is, and the only reason that the Sun is such an immense source of energy is because of its huge mass – even though only a small fraction of it is active in nuclear reactions.

Temperature

Astrophysicists need to be a bit more precise about what they mean by the temperature of a star. Remember that stars are good approximations to blackbodies and astrophysicists define a quantity called the **effective temperature** T_{eff} which is the temperature of a blackbody of the same size as the star that would emit the same total power. If we assume that all stars are spherical, then for a star of luminosity L, and radius R Stefan's Law (page 29) becomes:

$$L = 4\pi R^2 \sigma T_{eff}^4$$

It is important to realise that this is an estimate of the *surface temperature* of the star in the photosphere region. It is *not* the temperature of the interior of the star, which is quite different and determined by nuclear processes operating deep inside it.

The effective temperatures of stars range from very cool stars at about 2000 K, up to extremely hot ones of 100 000 K. The spectral characteristics of starlight provide information about temperatures in the stellar atmosphere and in Chapter 4 we saw how it is possible to assign a temperature to a star by classifying it by spectral type. Also, if we know its luminosity and radius, then we can use Stefan's Law to estimate its temperature.

Stellar radii

The Sun is the only star whose angular diameter can be directly observed and measured from Earth. By combining this with the mean distance between the Earth and the Sun, the Sun's true diameter can be calculated using simple trigonometry. All other stars are too far away for their angular diameters to be measured directly but stellar interferometry (page 78) can be used for the nearer ones. Table 5.3 shows some values of apparent angular diameters for five nearby stars.

Table 5.3 The diameters of five stars

Star	Apparent angular diameter/arcseconds	True diameter (Sun = 1)
Betelgeuse	0.034	730
Rigel	0.002 69	120
Canopus	0.006 86	82
Regulus	0.001 38	38
Sirius	0.006 12	1.9

Alternatively, if L and T_{eff} are known, then the radius of a star can be calculated by rearranging Stefan's Law to give:

$$R = \sqrt{\frac{L}{4\pi\sigma T^4}}$$

Stellar sizes range from 0.01 $R_\odot$ up to 1000 $R_\odot$. The star *Betelgeuse* in the constellation of *Orion* has a radius of 2.6×10^{11} m or approximately 2 AU. If *Betelgeuse* replaced the Sun in our Solar System then its radius would extend beyond the orbit of Mars!

Binary stars and stellar masses

In Chapter 1 we mentioned that at least 50% of all stars exist as pairs orbiting around each other in binary systems. Some of these 'double stars' can even be seen with the naked eye. One example is the double star system *Alcor–Mizar* in the constellation of *Ursa Major* or the Great Bear (Figure 5.1). A test of how good your eyesight would be to see if you can resolve the individual stars (you need a clear night to do this, preferably with no Moon). *Alcor–Mizar* constitutes a binary system with the two components revolving around their common centre of mass. They are separated by about 15 arcseconds and are so far apart that their period of revolution is several thousand years.

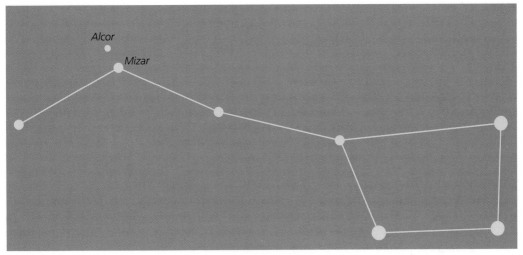

Figure 5.1 The position of the binary star system *Alcor–Mizar* in the constellation of *Ursa Major* or the Great Bear

The discovery of binary star systems was important evidence that Newton's Law of gravitation held outside the Solar System and gave astrophysicists reason to believe that the laws of physics could be applied universally. Systems in which two stars can be seen to orbit around each other in a definite period of time (typically tens of years) are called **visual binary** star systems.

Determining the masses of visual binaries

Binary stars offer astrophysicists the opportunity of measuring the mass of a star. By making observations of them through a telescope and using principles from astrodynamics, information can be gained about their respective masses.

Each star in a binary system rotates about the **centre of mass** of the system. This is the point in space where the total mass of the system appears to be concentrated and it is always closer to the more massive of the two stars (see Box 5.1).

Consider two stars with masses M_1 and M_2 that move in circular orbits with radii a_1 and a_2 about a centre of mass C (Figure 5.2).

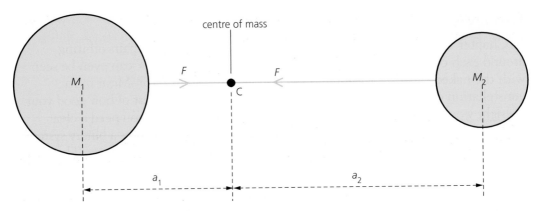

Figure 5.2 Centre of mass of a binary star system

By definition of the centre of mass (see Box 5.1):

$$M_1a_1 = M_2a_2$$

The force F acting between the two stars is gravity. Using Newton's Law of Gravitation, we can equate this to the centripetal force (see Box 2.2, page 43):

$$\frac{GM_1M_2}{(a_1 + a_2)^2} = M_1a_1\omega^2 = M_2a_2\omega^2$$

where ω is the common angular velocity of the two stars (in rad s^{-1}).

The ratio of the masses $M_1/M_2 = a_2/a_1$ and this can be rearranged to obtain $a_2 = (M_1/M_2)a_1$. So we can write:

$$a_1 + a_2 = a_1\left(1 + \frac{M_1}{M_2}\right)$$

and hence

$$a_1 = \frac{M_2(a_1 + a_2)}{(M_1 + M_2)}$$

From above

$$M_1a_1\omega^2 = \frac{GM_1M_2}{(a_1 + a_2)^2}$$

and substituting for a_1 we get:

$$\frac{M_1M_2(a_1 + a_2)\omega^2}{(M_1 + M_2)} = \frac{GM_1M_2}{(a_1 + a_2)^2}$$

If the time for the stars to complete one revolution of the binary orbit is T then $\omega = 2\pi/T$ so that:

$$\frac{T^2}{4\pi^2(a_1 + a_2)^3} = \frac{1}{G(M_1 + M_2)} = \text{constant}$$

This is Kepler's 3rd Law generalised to two masses. Thus if we know the period T and the separation of the stars $(a_1 + a_2)$ then we can obtain the sum of their masses. By observing the stars carefully the position of the centre of mass can be found to give the ratio M_1/M_2 and since $M_1 + M_2$ is also known the individual masses of the two stars can be calculated.

An accurate determination of the separation of the stars depends on knowing how far they are from the Earth. Sometimes this can be done for the nearer binaries using parallax methods or by using the distance–magnitude relation. In addition, the observer has to correct for the angle at which the orbit is inclined to the line of sight.

Box 5.1 Centre of mass

> The centre of mass of two bodies is the point at which their total mass appears to act.
>
> Consider two bodies M_1 and M_2 rotating in circular orbits around a common centre of mass C at distances a_1 and a_2 from C, as shown in Figure 5.2. Although the two bodies move at different speeds v_1 and v_2, their period of revolution T must be the same.
>
> As gravity acts between them the centripetal forces F_1 and F_2 are along the same line of action and we can write:
>
> $$F_1 = \frac{M_1 v_1^2}{a_1} = \frac{M_1}{a_1} \cdot \left(\frac{2\pi a_1}{T} \right)^2 = \frac{4\pi^2 a_1}{T^2} M_1$$
>
> and
>
> $$F_2 = \frac{M_2 v_2^2}{a_2} = \frac{M_2}{a_2} \cdot \left(\frac{2\pi a_2}{T} \right)^2 = \frac{4\pi^2 a_2}{T^2} M_2$$
>
> Newton's 3rd Law requires these forces to be equal and opposite so $F_1 = F_2$ and
>
> $$\frac{a_1}{a_2} = \frac{M_2}{M_1}$$
>
> which defines the position of the centre of mass.

Spectroscopic binaries

These are double star systems in which the orbital motion about the centre of mass is observed as a Doppler shift in the spectra of each star. Over a period of time, the set of spectral lines for each star shift back and forth and if at some instant one star is moving towards the Earth then its lines are blue-shifted whereas those of the other star (which is moving away) are red-shifted.

Using the Doppler relation

$$\frac{\lambda - \lambda_0}{\lambda_0} = \frac{v}{c}$$

the displacement of the spectral lines allows the star's **radial velocity** (or velocity along the line of sight) to be measured. A radial velocity curve can be constructed showing how the radial velocity varies with time (Figure 5.3). The *orbital* speeds v_1 and v_2 of each star can therefore be found and their orbital period T from the period of the radial velocity curve.

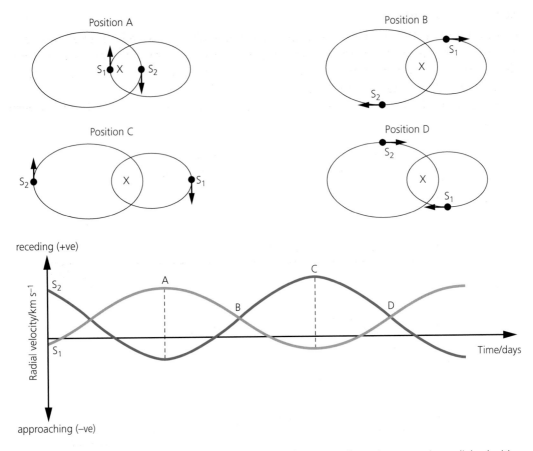

Figure 5.3 Binary stars and their radial velocity curves. The curves show the respective radial velocities of two stars S_1 and S_2 at the four points A, B, C and D of their orbits. X marks the position of their centre of mass

In the case where the orbits are circular, the circumference $2\pi a_1 = v_1 T$ and so:

$$a_1 = \frac{v_1 T}{2\pi} \quad \text{and} \quad a_2 = \frac{v_2 T}{2\pi}$$

and (from Box 5.1) we can write:

$$\frac{M_1}{M_2} = \frac{a_2}{a_1} = \frac{v_2}{v_1}$$

If the angular separation of the stars can be measured and their distance from the Earth is known, then $a_1 + a_2$ can be calculated.

Using Kepler's 3rd Law for two masses, $M_1 + M_2$ can be found, and knowing both the sum and the ratio of the masses the individual mass of each star can be calculated.

Eclipsing binaries

Useful information can also be determined if the double star system is an **eclipsing binary**, in which the stars are observed to pass in front of each other periodically. For this to happen, the orbit must be nearly edge-on as seen from the Earth. As one star eclipses the other, the apparent brightness of the binary image decreases and, by using photometry, the astrophysicist can measure the variation in light intensity with time (Figure 5.4).

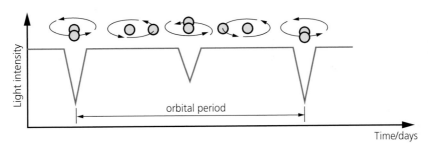

(a) Partial eclipse

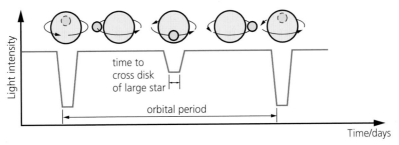

(b) Total eclipse

Figure 5.4 Light curves of eclipsing binaries

Light curves of eclipsing binaries can be used to obtain much detailed information about the nature of the two stars. The time between successive eclipses gives information about the orbital period and is therefore related to the separation of the stars by Kepler's 3rd Law; and if the eclipsing binary is also a spectroscopic one, the individual velocities of the stars can be calculated, which allows the masses of the two stars to be determined.

The duration of the eclipse depends on the size and the speed at which the eclipsing star travels. These observations are used to estimate the diameter of the star from geometrical considerations and the shape of the light curve. Also, by observing precisely how the light from the eclipsed star is cut off, astrophysicists can deduce information about the star's atmosphere including its temperature and pressure. To date astronomers have discovered about 4000 eclipsing binary star systems but only a fraction of these have light curves which are good enough to interpret accurately, due to the weak amount of light detected and the spectrum being too faint for radial velocity measurements to be made.

Chemical composition and age

In Chapter 4 we discussed the kind of information that can be deduced from the spectrum of a star. The spectral line patterns of a star reveal the 'signature' of the chemical elements which are present and allow us to deduce the chemical composition of the star and fit it into a spectral classification scheme.

Most stars are very similar in their composition. By mass about 72% is hydrogen and 25% helium and the remaining elements make up only 3% of the mass with their relative abundances being more or less the same. These elements include O, C, N, Si, Fe, Mg, S, Al, Ca, Na and Ni, and are what astrophysicists loosely call 'metals' even though they include obvious non-metals such as carbon, nitrogen and oxygen! Some stars though do not fit into the classification scheme because they are 'metal poor' with much smaller proportions of the same elements although in similar relative abundances. To explain this, astrophysicists broadly divide stars into two **populations**:

Population I stars   These are stars like the Sun, whose elemental composition is well defined and can be fitted into the spectral classification scheme. The hottest of these stars are blue, and are generally found in the spiral arms of galaxies. They are also associated with interstellar gas clouds. These stars are relatively young compared with the age of their galaxies and were formed from an interstellar medium already enriched with heavier elements made by nuclear reactions inside stars from previous generations of stellar life and death.

Population II stars   These are metal poor and the brightest ones are red. They tend to be found outside spiral galaxies in globular clusters at high inclinations to the galactic disc. They are very old stars, born soon after the formation of a galaxy before many heavier elements had been synthesised by succeeding stellar generations. As a result, they are rich in hydrogen and helium but contain only smaller proportions of the heavier metals.

It is not possible to follow the lifespan of a star by observation. Even for relatively short-lived stars the timescales are impossibly long, being many thousands of years compared to hundreds of millions of years for the longest-lived ones. In order to determine the age of a star we must use indirect methods based on theoretical models and compare it with others on the basis of physical similarities.

The Hertzsprung–Russell Diagram

Suppose we plot a graph of magnitude versus spectral type for all types of stars for which these quantities can be measured. Then we obtain a diagram like the one illustrated in Figure 5.5a, which is known as a **Hertzsprung–Russell Diagram (HRD)** after Ejnar Hertzsprung (1873–1967) and Henry N. Russell (1877–1957), two astronomers who first made this kind of plot.

Since magnitude is a measure of a star's luminosity, and spectral type is related to how hot a star is, the HRD is essentially a plot of the luminosity of stars against their temperature. Remember in Chapter 2 we said that stars are very good approximations to blackbodies. This means that their colour is related to their temperature. Figure 5.5b shows the HRD in terms of a star's colour as a function of its surface temperature and luminosity. A great deal of information about the attributes of stars can be obtained from it, especially how they evolve with time.

If you look at Figure 5.5a you will see that the points on the graph are not randomly scattered but are divided into several distinct groupings of stars. Also notice that the temperature axis is scaled so that it increases *from right to left*, i.e. the hotter stars are towards the left of the diagram.

The band stretching diagonally across from the top left to the bottom right is called the **main sequence** and comprises most of the stars we can see at night. At the top of the main sequence are the hot luminous blue stars and at the bottom cool, dim reddish stars. Any star whose luminosity and temperature place it in this band is called a **main-sequence star**. The Sun is a main-sequence star and its position is shown on the diagram.

In the upper right-hand side of the HRD is a second grouping of stars. These are stars that are both luminous and cool; they are **red giants** which are luminous because of their very large size despite their surface temperatures being relatively low, in the range 3000–4000 K. The smattering of stars towards the top right-hand corner which are larger and brighter than the red giants are the **supergiants**. Supergiant stars have very large diameters and therefore high luminosities, but again low surface temperatures.

The final group is towards the bottom left-hand corner of the HRD. These are **white dwarfs**, stars that are both hot and dim. They are very small – about the size of planets. Although they have very high temperatures, white dwarfs appear dim because of their small size and hence small area of radiating surface.

The significance of the HRD is that it tells us that there exist fundamentally different kinds of stars. 'Normal stars' like the Sun are those that lie along the main sequence. 'Unusual stars' are the giants and white dwarfs which seem to have a very different relationship between luminosity and temperature. In the night sky about 90% of stars belong to the main sequence. The rest is divided between the red giants, white dwarfs and some odd varieties. In Chapter 8 we will see how the HRD actually represents different stages of stellar evolution – how stars are born, grow old and then die.

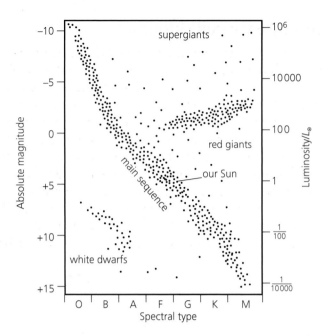

Figure 5.5 (a) The Hertzsprung–Russell Diagram (HRD). The HRD shows us that there are different types of stars

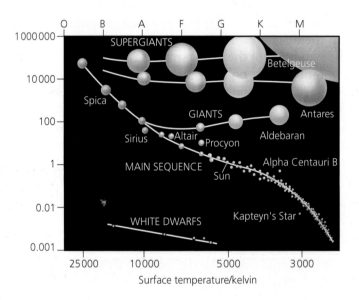

Figure 5.5 (b) A true colour representation of the Hertzsprung–Russell diagram. The horizontal axis is temperature (along the bottom) and spectral type (along the top); the vertical one is luminosity, in units of the Sun's luminosity. The solid white lines show where stars of different luminosity classes fall on the diagram: 1a supergiants at the very top, 1b supergiants below these, III giants just below the supergiants, and finally luminosity class V main-sequence stars. The relative sizes of the stars are shown correctly within each luminosity class, but not between them. The colours are those as perceived by the eye looking at these stars through a telescope

The mass–luminosity relation

Let's return to the main-sequence stars again. How does one main-sequence star differ from another? Careful observations over many years of the orbits of binary stars whose distances are known have provided astrophysicists with the masses of many main-sequence stars. Figure 5.6 shows a graph of luminosity versus mass for stars on the main sequence. You can see that there is a definite correlation between mass and luminosity which shows that the more massive the star then the more luminous it is. This connection is known as the **mass–luminosity relation** and it shows that on the Hertzsprung–Russell Diagram, the stars in the main sequence follow a progression in *mass* as well as in luminosity and surface temperature. The hot blue stars at the top left of the main sequence are the most massive ones, while the cooler reddish ones at the bottom right are the least massive and it therefore follows that stars of intermediate luminosity and temperature on the main sequence also have intermediate mass.

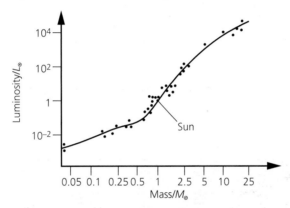

Figure 5.6 The mass–luminosity relation. For main sequence stars this shows a definite correlation between luminosity and mass

Summary

◆ A star is characterised by six physical properties: **luminosity, temperature, radius, mass, chemical composition** and **age**. By careful observation of stars using a number of methods, astronomers and astrophysicists can determine these properties either experimentally or from theory.
◆ Binary systems are important in measuring a star's physical properties. There are three types: **visual, spectroscopic** and **eclipsing** binaries. From binary stars we can determine stellar masses and, in the case of eclipsing binaries, information about the temperature and atmosphere of a star.
◆ The **Hertzsprung–Russell Diagram** is a graph of luminosity versus temperature and shows that stars can be grouped together on the basis of their physical similarities. Most stars on the HRD lie on a diagonal line called the **main sequence**.
◆ The **mass–luminosity relation** shows that the more massive a star the more luminous it is and from this we can infer that stars on the main sequence differ from each other by reason of their mass.

Questions

1 What are the six principal physical properties that characterise a star?

2 Astronomers have discovered from their observations of the star *Capella* that:

- its surface temperature T is 5200 K
- its distance from the Earth is 4.3×10^{17} m
- at the Earth's surface the intensity of the radiation received from *Capella* is 1.2×10^{-8} W m^{-2}.

Explain briefly how the surface temperature is determined.
Calculate the radius r of *Capella*, given that its luminosity can be found by using

$$L = 4\pi r^2 \sigma T^4$$

where σ is the Stefan–Boltzmann constant which is 5.7×10^{-8} W m^{-2} K^{-4}.

ULEAC, Jan 1997, part

3 The diagram below shows a pair of stars, X and Y, a distance r apart forming a binary system. Their angular speed about their centre of mass C is ω. Their masses and their distances from C are m_1, r_1 and m_2, r_2 respectively.

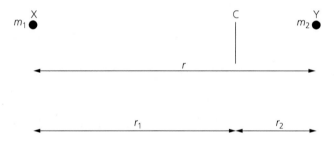

The following diagram shows how the brightness of the binary pair, seen as an unresolved single star, varies with time.

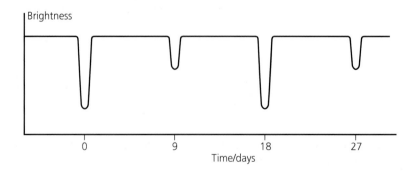

Show that the angular speed ω of the binary pair is 4.04×10^{-6} rad s^{-1}.

The binary pair is viewed through a filter which transmits light of wavelength close to the helium line 447 nm. The diagram below shows how the wavelength λ of the light from the binary pair for this helium line changes during one full period of revolution. There are two curves because there are two stars.

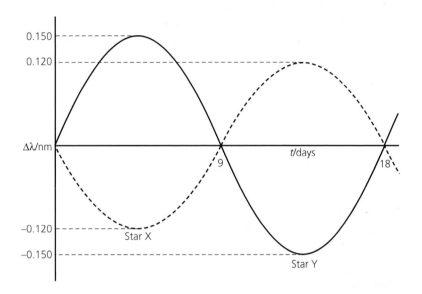

How do you account for the shape of the dashed curve? Explain why the two curves are exactly out of phase.

Calculate the maximum forward speed of star X and the values of r_1, r_2 and r.

ULEAC, Jan 1998, part

4 What is the mass–luminosity relation? What is its connection with the Hertzsprung–Russell Diagram?

5 Show that, in the case of a planet orbiting the Sun, Kepler's 3rd Law:

$$\frac{T^2}{4\pi^2(a_1 + a_2)^3} = \frac{1}{G(M_1 + M_2)}$$

reduces to:

$$\frac{T^2}{a^3} = \frac{4\pi^2}{GM_\odot} = \text{constant}$$

by taking $M_1 = M_\odot$ the mass of the Sun, $M_2 =$ the mass of the planet ($M_1 \gg M_2$) and $a_1 + a_2 = a$ the distance of the planet from the Sun.

Star formation and the interstellar medium

Observations of the interstellar medium, together with theoretical stellar models, have led astrophysicists to develop ideas on how stars are formed and why they shine. The interstellar medium contains the raw materials out of which stars are made and the galaxy contains giant interstellar nurseries where stars are being born. Thanks to instruments like the Hubble Space Telescope, we can look into these nurseries and see the process of starbirth in action.

The interstellar medium

The space between the stars is not empty. Figure 6.1 shows a picture of the *Eagle Nebula* taken by the Hubble Space Telescope. This is a huge cloud of gas and dust in the constellation of *Serpens* about 6500 ly from Earth. This remarkable picture shows protostars forming out of interstellar matter. What is this matter and how does the process of star formation come about? To answer these questions we must first of all take a closer look at the **interstellar medium (ISM)** to see what it contains.

Figure 6.1 The *Eagle Nebula*. These dark pillar-like structures are clouds of cool interstellar hydrogen gas and dust in which stars are being born. The tallest pillar (left) is about a light year from base to tip

The interstellar medium is found to be composed of the following:

1 clouds of interstellar gas in the form of neutral and ionised hydrogen
2 a hot 'intercloud medium' of thin hydrogen gas, some of which is at very high temperatures
3 very large 'molecular clouds' of organic and inorganic molecules
4 interstellar dust.

The ISM is a complex mixture of atoms, molecules and dust, and to understand how it all fits together let's look first at interstellar gas and then at interstellar dust, so that we can use the observational evidence to try to develop a plausible theory of star formation.

Interstellar gas

At the beginning of the 20th century, a German astronomer Johannes F. Hartmann (1865–1936) showed that the interstellar void was filled with a tenuous gas at very low density. Hartmann was interested in the spectra of binary stars and he noticed that the spectra of some exhibited two absorption lines whose wavelengths *did not* change despite the periodic variation of all their other spectral lines. He found that these fixed lines belonged to ionised calcium and deduced that they were formed not in the outer layers of the binary stars, but somewhere in the line of sight of the observer. This surprising result was how interstellar gas was first discovered – starlight was passing through the gas and being absorbed in certain narrow regions of the visible spectrum.

Most of the ISM consists of atoms of hydrogen created very early on in the history of the universe (see Chapter 10); a much smaller proportion consists of other elements formed in the final phases of the life cycle of stars and ejected in supernova events or blown off the surfaces of giant stars. The hydrogen gas in the ISM tends to clump together in clouds, but the average density of interstellar gas is very small – there are only about 10^6 hydrogen atoms per cubic metre, which is a very low vacuum by our laboratory standards!

Emission nebulae

As well as Hartmann's discoveries, we know that gas exists between the stars because of **emission nebulae**. Figure 6.2 shows a picture of the *Orion Nebula*, about 1500 ly from Earth in the winter constellation *Orion*. On a clear night, just below the three stars of Orion's belt a fuzzy patch of light can be seen. A small telescope or pair of binoculars will easily resolve this as a glowing, diffuse cloud of gas surrounding a small cluster of stars.

When we look at the spectrum of the *Orion Nebula*, we find that it contains bright emission lines of hydrogen, helium and oxygen. Emission nebulae do not radiate light by themselves but do so due to fluorescence. Atoms in the gas absorb ultraviolet photons from hot (O spectral type) stars which are near or embedded in the nebula, and as they de-excite they emit photons at lower energies, for example the red

Figure 6.2 The *Orion Nebula*. This is a hot emission nebula. The bright central region is about 16 ly across

Balmer line at 656.5 nm which gives the emission nebula a reddish appearance. The hot stars heat the gas to temperatures greater than 10 000 K and as a result much of the hydrogen in the nebula is ionised. Such a volume of ionised hydrogen is called an **H II region**. (The abbreviation 'H II' means hydrogen that is ionised whereas 'H I' stands for neutral hydrogen.) The H II region in *Orion* extends for many tens of parsecs, beyond which is an H I region of neutral hydrogen gas.

By mass, H II regions form less than a tenth of the total mass of the ISM. These regions also fluoresce at radio wavelengths with a continuous spectrum which allows radio astronomers to map their size and infer how much ionised hydrogen is present.

Clouds of neutral hydrogen gas

H I regions are much cooler than H II regions, about 50–100 K, and almost all of the hydrogen atoms are in the ground state. As well as existing on the perimeters of H II regions, neutral hydrogen atoms also clump together in smaller clouds a few parsecs across. There is much more H I than H II gas in the ISM, in fact some 40% of the total mass of the ISM consists of H I regions. Since the temperature of H I gas is low and its electrons are in the ground state, it does not emit radiation by fluorescence and the atoms do not have sufficient kinetic energy to be excited by collisions. However, there is one way in which they can be detected at radio wavelengths. Hydrogen atoms have one electron orbiting a single proton nucleus. As well as the electron's motion around the nucleus, quantum mechanics says that the electron has a property called **spin**, which is a measure of rotational motion about its own axis. The proton also has spin. In a hydrogen atom, the proton and electron can be regarded as miniature tops which can be orientated with their spins either in the same direction or in opposite directions. Each of these states has a different total energy with one higher than the other. Occasionally, the electron 'flips' from the higher energy to the lower energy state and a radio frequency photon is emitted, with a wavelength of 21.11 cm (see Box 6.1).

This **21 cm radiation**, as astrophysicists call it, was first detected in 1951. It can be used to map out the distribution of neutral hydrogen in the Milky Way. Surveys at 21 cm using radio telescopes have found that most neutral hydrogen is located in the plane of the spiral galaxy, with an average temperature of 70 K and a density of 5×10^5 atoms m^{-3}.

Box 6.1 The 21 cm line

H I regions contain neutral hydrogen atoms which can be detected by their radio emissions using radio telescopes tuned to a wavelength of 21 cm. How does this emission process occur?

The hydrogen atom consists of a single proton with a single electron in orbit around it. Now both particles can be regarded as spinning like miniature tops and therefore have both angular momentum and rotational energy. The laws of quantum mechanics tell us that the spins of the proton and electron can *only* be orientated either so that their spins are in alignment, or in the opposite sense. No 'in between' orientations are allowed. If the two spins are opposed, then the total energy of the atom is slightly less than if they are aligned. If initially the spins are aligned then the electron spin can flip over (or be reversed) and the atom will drop to a lower energy state. This results in the emission of a photon.

The energy difference between these two states corresponds to a frequency of about 1427 MHz or a wavelength of about 21 cm. This is in the radio region of the spectrum.

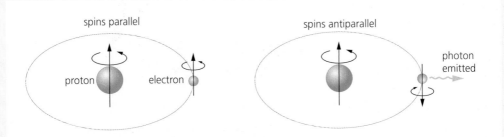

Figure 6.3 The 21 cm hydrogen line is produced by the electron in a hydrogen atom flipping from one spin state to the other

The probability of a flip taking place spontaneously is very small; on average about 10^7 years must elapse before it happens, but because there are so many hydrogen atoms in interstellar space, there will always be enough that are emitting 21 cm photons for this wavelength to be strongly detected in the interstellar gas.

Intercloud gas

Between the clouds of H II and H I regions is a hot, tenuous **intercloud gas** composed of neutral and ionised hydrogen. This consists of very hot coronal gas of ionised hydrogen at temperatures as high as 10^6 K but a low density of 100 to 1000 atoms m^{-3}, and cooler hydrogen between 7000 K and 10 000 K at a higher density of 10^5 atoms m^{-3}. It is estimated that the intercloud gas constitutes about 20% by mass of the ISM. The hot coronal gas was discovered comparatively recently, in 1976, by the *Copernicus* space astronomy satellite, from the X-rays it emits.

Molecular clouds

Using spectroscopy, astronomers made the first discoveries of interstellar molecules in space by analysing absorption lines in interstellar spectra. The energy level structure of molecules is very complex and the vibrational and rotational states of a molecule tend to cause absorption or emission of radio photons at millimetre wavelengths. Using radio telescopes, many different organic molecules (containing carbon) and inorganic molecules have been discovered. The most abundant is molecular hydrogen H_2.

Interstellar molecules are concentrated in dark, dense, cold clouds, at temperatures of 10–50 K and densities of 10^8–10^{15} molecules m^{-3}. The size of these molecular clouds varies from 50 to 100 ly across and the densities are so high that gas mixed with interstellar dust blocks starlight from behind, causing them to appear as dark patches in an otherwise bright starfield. Figure 6.4 shows a molecular cloud region called the *Coal Sack* in the constellation *Crux*.

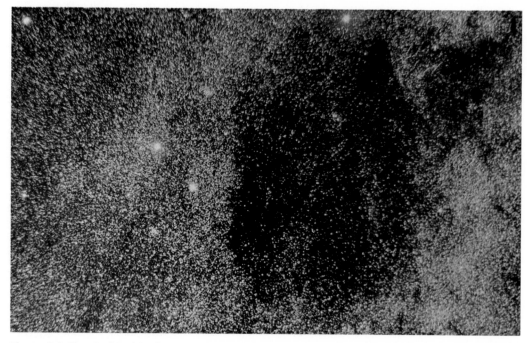

Figure 6.4 The *Coal Sack* in the constellation *Crux*. This is a molecular cloud region from which stars can form

Molecular clouds tend to be held together in equilibrium by their own gravity; the random motions of the gas molecules, which would tend to make the cloud disperse, are counteracted by their own gravitational attraction. They have an average mass of about $1000 M_\odot$ and make up some 40% of the total mass of the ISM. H II regions are often found near molecular clouds, and the *Orion Nebula* is a good example. It is adjacent to a molecular cloud some 33 ly across with a mass of $10^4 M_\odot$.

Significantly, astronomers observe that H II regions are often found near young massive stars and we will see that this fact strongly suggests that molecular clouds have something to do with the process of star formation.

Interstellar dust

As well as gas the Milky Way contains vast lanes of dust which do not shine. It was originally thought that these dark areas were due to the absence of stars. In fact, **interstellar dust** forms about 1% by mass of all interstellar matter and being able to see through it has now become possible thanks to advances in radio and infrared astronomy. This has revealed important new insights into the process of star formation.

The presence of dust is easily seen in Figure 6.5 which shows a dust cloud in the constellation *Orion*. It is called the *Horsehead Nebula* owing to its characteristic shape and is an example of a **dark nebula**. The dust obscures the light from distant objects as well as from those inside it.

Figure 6.5 The *Horsehead Nebula*. This is a dark nebula of dust grains which blocks light from the background emission nebula

Figure 6.6 An example of a reflection nebula in the star cluster *Pleiades*. The cloud of dust glows with reflected light

Another way in which interstellar dust is revealed visually is when it reflects starlight. Figure 6.6 shows a picture of the Pleiades star cluster surrounded by a nebulous cloud of glowing dust. When analysed, the spectrum of the nebula is found to be simply that of the reflected spectrum of the *Pleiades* stars. The dust is illuminated by the stars in the cluster and reflects their light towards us. Bright nebulae which shine in this way are called **reflection nebulas**.

Detection of dust

Interstellar dust can be detected due to four effects that it has on starlight: **extinction**, **reddening**, **polarisation** and **infrared emission.**

We mentioned in Chapter 3 that the effect of interstellar matter needs to be taken into account when determining the apparent magnitude of a celestial object. **Extinction** is the dimming of starlight as it travels through the dust. The dust particles either absorb some of the light or scatter it, so that less light emerges from the dust than went in.

Extinction depends on the wavelength of the incident light – blue light is more strongly scattered and absorbed than red. Red light will therefore pierce dust more easily than blue so when you observe a star through a dust cloud more light at red wavelengths reaches your eye. The star therefore appears redder than it actually is and astrophysicists call this effect **reddening**.

Starlight can become **polarised** when passing through a dust cloud if the dust particles are orientated in a preferred direction. The alignment of the dust is believed to be caused by magnetic fields (the particles lining up with the direction of the field), so the detection of polarised light is evidence of the existence of **interstellar magnetic fields**. Using polarimeters, studies have shown that the interstellar magnetic field strength is 10^{-10} T which is about 100 000 times less than the surface magnetic field of the Earth.

Finally, the dust particles emit **infrared radiation**. Figure 6.7 shows an image taken by the Infrared Astronomy Satellite (IRAS). IRAS detected infrared radiation coming from all directions in space and showed that infrared emission was the strongest in regions where there is a high concentration of interstellar gas and dust. The emission is called **infrared cirrus** due to its resemblance to the high altitude cirrus clouds in the Earth's atmosphere. The grains of interstellar dust emit infrared radiation because they behave like very small blackbody radiators. Each grain will absorb light from a nearby star and heat up until it emits, in the infrared, as much energy per second as it absorbs. It has then reached thermal equilibrium. The temperature of interstellar space can range from 3 K to 3000 K, and Wien's Law tells us that the peak wavelength of emission will be in the infrared band of 1000 μm to 1 μm.

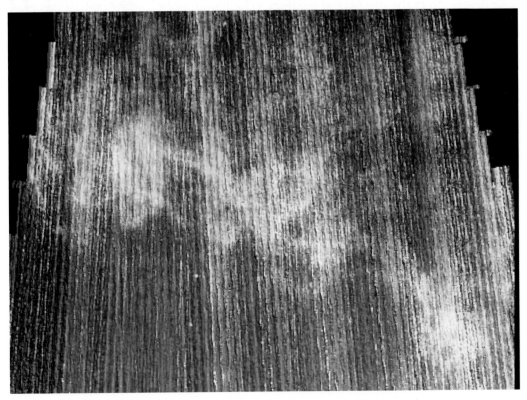

Figure 6.7 Infrared cirrus. This IRAS image shows infrared emission from interstellar dust. (The vertical stripes are artefacts produced by IRAS's scanning infrared detector)

Composition of dust

From extinction, astrophysicists are able to estimate the size of the interstellar dust grains. They are found to range from 0.02 µm to 0.5 µm. From studies of their absorption and emission spectra, it is thought that interstellar dust contains elements of H, C, O, Si, Mg and Fe in the form of ices, silicates, graphite and metals.

Interstellar dust originates from the atmospheres of cool giant and supergiant stars. These stars have surface temperatures of only about 2500 K and their stellar winds blow mass into space at about 10^{-6} solar masses per year. As we shall see later in Chapter 9, an important dust source is material that has been ejected from nova and supernova explosions.

Interstellar dust has a role to play in the formation of molecules. Ultraviolet radiation from nearby stars can provide the necessary energy for combining atoms in order to make molecules but, by the same token, can dissociate them into their constituent atoms. The dust can act as an absorber of some ultraviolet radiation, shielding the newly formed molecules from dissociation.

Starbirth

Our discussion of the interstellar medium so far has all been leading up to a theory of star formation and the main idea is this:

Stars are born in the interstellar medium by the gravitational collapse of gas and dust within interstellar molecular clouds which have mass many times greater than the mass of a single star.

What do we mean by gravitational collapse? Well, the clouds condense into small bodies under the action of gravity. The exact details of this process are still far from understood, but by combining theoretical ideas together with observational evidence, astrophysicists can develop a theory of how stars are formed.

There are two main theoretical ideas to consider. These are **gravitational instability** and **free-fall gravitational collapse**.

Gravitational instability

What causes the clouds to become unstable and condense in the first place? This problem was first investigated by a British astrophysicist Sir James Jeans (1877–1946). He asked the question *how big a mass of gas is necessary to make sure that it will collapse under its own self-gravity?*

To simplify matters let's start with an isolated spherical mass of gas of uniform density. Three main factors – the **mass**, the **density** and the **temperature** of the gas – will determine whether the cloud will condense or not. The strength of the gravitational attraction on a particle of gas will clearly depend on the total mass of the cloud. The size of the cloud will also be important because of the inverse square law of attraction and the greater the density of the cloud, the greater will be the attraction between the individual particles.

The motion of the gas particles is random. Their kinetic energy will be determined by the temperature of the gas and this has the effect of dispersing the particles in the cloud, so opposing their tendency to attract each other.

Jeans realised that there must be a critical value for a mass of gas such that, above a certain limit, gravity would overcome the thermal motion of the particles and the cloud would condense; and below the limit, gravity would be too weak and the thermal motion of the particles would disperse the cloud. This critical mass of gas is called the **Jeans mass** (Box 6.2).

The reality, though, is more complicated than this. We have assumed a spherical gas cloud of uniform density whereas in fact it may be uneven, giving rise to more than one centre of attraction. This means that more than one star could form within the cloud and this is what we observe. Also, observations show that interstellar gas clouds are irregularly shaped. Furthermore, the cloud will be rotating so angular momentum will also play a part in the condensation. We know also from observations of, for example, the *Orion Nebula* that the interiors of interstellar clouds are turbulent and the gas is not in uniform motion. Despite these complications, the Jeans theory provides a simplified starting point in trying to explain the very complex process of starbirth.

Box 6.2 The Jeans mass

The **Jeans mass** is the critical mass of gas needed for an interstellar gas cloud to condense under its own gravity.

Let's consider a spherical mass of gas M of uniform density ρ with radius R. A gas molecule of mass m on the perimeter of the cloud which (because of its thermal motion) has speed v, will have kinetic energy $\frac{1}{2}mv^2$.

The molecule will also have **gravitational potential energy** of $-GMm/R$. The magnitude of this quantity is the amount of energy needed to remove the molecule completely from the cloud. This implies that (since the molecule is on the perimeter):

$$\frac{GMm}{R} > \frac{1}{2}mv^2$$

or

$$\frac{2GM}{R} > v^2$$

Now the temperature T of the molecules is a measure of their average kinetic energy and may be expressed as

$$KE \text{ (average)} = \frac{1}{2}mv^2 = \frac{3}{2}kT$$

where k is Boltzmann's constant. Rearranging, we see that:

$$v^2 = \frac{3kT}{m}$$

(continued)

If we substitute for v^2 in the equation above, we obtain:

$$\frac{2GM}{R} > \frac{3kT}{m} \quad \text{or} \quad \frac{2GMm}{3kT} > R$$

We will see that it is helpful to express this as:

$$\left[\frac{2GMm}{3kT}\right]^3 > R^3$$

Now the mass of the cloud is equal to its density multiplied by its volume and so:

$$M = \tfrac{4}{3}\pi R^3 \rho \quad \text{or} \quad R^3 = \frac{3M}{4\pi\rho}$$

Substituting for R^3 we can write:

$$\left(\frac{2Gm}{3kT}\right)^3 M^3 > \frac{3M}{4\pi\rho}$$

$$\therefore M^2 > \left(\frac{3}{4\pi\rho}\right) \times \left(\frac{3kT}{2Gm}\right)^3$$

$$> \left(\frac{81}{32\pi\rho}\right) \times \left(\frac{kT}{Gm}\right)^3$$

$$\therefore M > \frac{9}{4(2\pi\rho)^{1/2}} \times \left(\frac{kT}{Gm}\right)^{3/2}$$

What does this last expression tell us? It says that a spherical gas cloud of density ρ and temperature T will be gravitationally held together if its mass is equal to (or greater than) the right-hand side of the equation. If T is reduced then the internal gas pressure gets smaller and the gravitational force pulls the gas molecules closer together. If, however, the mass is less than this limit then the thermal motion of the molecules will overcome gravity and the cloud will disperse. This critical mass is called the Jeans mass M_J:

$$M_J = \frac{9}{4(2\pi\rho)^{1/2}} \times \left(\frac{kT}{Gm}\right)^{3/2}$$

Gravitational collapse

How long does it take for a cloud to condense to form a star? We assume that the minimum time for this to happen is when the cloud contracts entirely under the influence of gravity and we neglect any internal pressure resisting the collapse. This is called the **free-fall collapse time** and, as Box 6.3 shows by applying Kepler's 3rd Law, it depends only on the cloud's *initial density* and not its mass.

Once the cloud has acquired enough mass to overcome its own outward internal pressure and start to contract, then gravity effectively becomes the only force acting on the gas particles and they collapse in free fall. Some of the dust particles collide with each other as they fall towards the centre where their gravitational potential energy is converted into random kinetic or thermal energy of the particles.

So long as the cloud is mostly transparent to infrared radiation, this thermal energy is radiated into space, and the cloud remains relatively cool and its internal pressure low. The collapsing cloud at this stage is called a **protostar**.

As the free fall continues, the density of the cloud increases and it becomes opaque. The thermal energy becomes trapped and the cloud starts to heat up, thereby increasing its internal pressure. An important first stage in the formation of the star is then reached. The temperature (and density) of the cloud have reached a level where the force due to internal pressure resists the gravitational collapse, and the tremendous heat generated causes the protostar to become luminous. It is then called a **pre-main sequence star**.

Pre-main sequence stars radiate heat and light only weakly. In fact the process of starbirth is hidden from our direct view and 'stellar nurseries' can only be observed using infrared and radio observations.

Stars like the Sun, however, are clearly visible and radiate with intense heat and light. So far we have said nothing about how this huge amount of energy is generated. To answer this question in the next chapter we will see that the source of stellar energy lies in the nuclear reactions initiated when a pre-main sequence star becomes hot enough to 'switch on' and shine.

Box 6.3 Gravitational collapse and free-fall time

Consider a spherical cloud of gas and dust with radius r. Imagine a particle of mass m on the edge of the cloud. Now gravity acts in the cloud as if all its mass were concentrated at the centre. Let the total mass of the cloud be M and its initial average density ρ_0. The particle will fall towards the centre along a radius of the sphere. We can use Kepler's 3rd Law to describe this motion if we think of the path as the orbit of an ellipse that is so flat that we can regard it as a straight line with the foci at opposite ends of the major axis, $2a$.

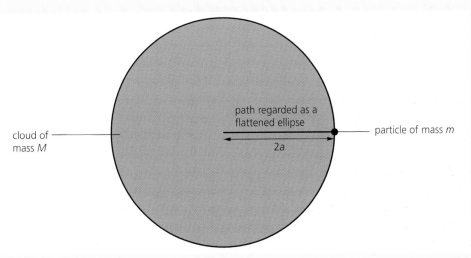

cloud of mass M

path regarded as a flattened ellipse

$2a$

particle of mass m

Figure 6.8 Modelling the motion of a particle at the edge of a gas cloud

(continued)

Using Kepler's 3rd Law (page 38) we can write:

$$\frac{T^2}{a^3} = \frac{4\pi^2}{G(M + m)} \approx \frac{4\pi^2}{GM} \qquad \text{since } M \gg m$$

and in this case, since the major axis $2a$ of the orbit is equal to r and mass = volume × density,

$$M = \tfrac{4}{3}\pi r^3 \rho_0 = \tfrac{4}{3}\pi(2a)^3 \rho_0 = \frac{32\pi a^3}{3}\rho_0$$

and so we can write Kepler's 3rd Law as:

$$\frac{T^2}{a^3} = \frac{4\pi^2}{G[(32\pi/3)a^3\rho_0]}$$

$$T = \left[\frac{4\pi}{(32/3)G\rho_0}\right]^{1/2}$$

However, T is one whole orbital period, i.e. the time it takes to go a distance r and r back again. The free-fall time is the time it takes for the particle to fall only the distance r and is therefore equal to *half* the period, or $T/2$. Therefore the time for free fall t_{ff} is given by:

$$t_{ff} = \frac{T}{2} = \left[\frac{\pi}{(32/3)G\rho_0}\right]^{1/2} = \left[\frac{3\pi}{32G\rho_0}\right]^{1/2}$$

Notice that the free-fall time depends *only* on the cloud's initial average density and not on its mass. If ρ_0 is in units of kg m^{-3} then

$$t_{ff} = \frac{6.65 \times 10^4}{\rho_0^{1/2}} \text{ seconds} = \frac{2.11 \times 10^{-3}}{\rho_0^{1/2}} \text{ years}$$

So how long does it take for a molecular cloud to collapse? Consider a giant molecular cloud containing a number of particles $n = 10^{12}$ m^{-3}. Now the mass of hydrogen m_p is 1.67×10^{-27} kg so that the density of the cloud is:

$$\rho_0 = nm_p = 10^{12} \times 1.67 \times 10^{-27} \approx 2 \times 10^{-15} \text{ kg m}^{-3}$$

$$\therefore t_{ff} = \frac{2.11 \times 10^{-3}}{\rho_0^{1/2}} = \frac{2.11 \times 10^{-3}}{(2 \times 10^{-15})^{1/2}} = 5 \times 10^4 \text{ years}$$

and this is a very short time compared to the average lifetime of a star.

Summary

◆ The space between the stars is called the **interstellar medium (ISM)** and is filled with gas and dust. The gas is mostly hydrogen in the form of atoms, molecules and ions. Regions of ionised hydrogen gas are called **H II regions**, and regions of neutral hydrogen **H I regions**. The neutral hydrogen can be detected and mapped by its radio emission at 21 cm.

◆ Using optical and radio observations, astronomers have found a variety of organic and inorganic molecules in space. Molecular hydrogen gas clumps together in clouds ranging in size from small to giant molecular clouds tens of parsecs across.

◆ Interstellar dust is less common than interstellar gas and forms about 1% of the total mass of interstellar matter. Dust can be detected by **extinction, reddening, polarisation** and **infrared emission**.

◆ Interstellar dust contains tiny grains a few μm in size consisting of silicates, graphite or iron. Dust clouds are associated with molecular clouds and assist in the formation of molecules. They mainly originate from the outflow of matter from cool supergiant stars and the residues of nova and supernova events.

◆ Stars are born by molecular clouds collapsing under their own self-gravity. Two important theoretical ideas are **gravitational instability** and **free-fall gravitational collapse**. When a star is undergoing free-fall collapse it is called a **protostar**.

◆ As the density and temperature of a protostar increase, a stage is reached when the collapse is slowed down due to the outward internal pressure of the cloud balancing gravity. In this stage, when gravitational energy is its only source of heat and radiation, the inaugural star is called a **pre-main sequence star**.

Questions

1 **a** What are the three main ingredients of the interstellar medium?

  b How do astronomers deduce the distribution of hydrogen in the Milky Way?

2 Assuming that an interstellar dust grain radiates as a blackbody and has an equilibrium temperature of 30 K, show using Wien's Law that it radiates in the far infrared end of the electromagnetic spectrum.

3 Suppose there is one dust particle per 10^6 m^3 and 0.33 hydrogen atoms per cm^3. If interstellar dust accounts for 1% of the total mass of interstellar matter, how massive must a dust particle be? (mass of H atom = 1.7×10^{-27} kg)
  If it has a density of 3000 kg m^{-3} what is its radius (assuming it is spherical)?

4 An H I region has a density of 10^9 atoms m^{-3} and is roughly spherical with a radius of about 1 pc. Estimate how much energy is required to completely ionise the nebula. (ionisation energy of H = 13.6 eV)

5 a A molecular cloud has a density of 10^{18} particles per cubic metre. Calculate its free-fall collapse time using the equation derived in Box 6.3.

  b The Sun is currently in a state of equilibrium – the outward force due to its internal pressure balances the inward force due to gravity. Suppose its internal pressure was suddenly removed. What would be the free-fall collapse time of the Sun? (average density of Sun = 1400 kg m^{-3})

6 Pogson's Law may be written as $m - M = 5 \log d - 5 + \alpha d$

where α is the interstellar absorption in magnitudes per parsec. For lines of sight along the plane of the galaxy, typical values of α are 0.002 mag pc^{-1}. Compare the apparent magnitude of a star with absolute magnitude –7.1 which at a distance of 276 pc when

  a the interstellar absorption is neglected

  b the interstellar absorption is taken into account.

Energy generation in stars

Knowledge of the atom and of nuclear theory has enabled astrophysicists to discover the source of stellar energy. The huge amounts of energy generated by stars are due to nuclear fusion reactions occurring deep within their interiors. Using physical assumptions and observational data, a model of a star can be made which describes how energy is transported from its interior into space.

The mass–energy relation

In Chapter 6, we looked at how protostars condense out of giant molecular clouds of gas and dust. Young stars are radiating light weakly, but their internal temperatures are not high enough for nuclear reactions to begin. To understand how stars can shine brightly, we must examine the structure of the atom and, in particular, the properties of the nucleus.

In the 19th century, evidence from geology and the fossil record indicated that the Earth must have been more or less in its present form for several hundred million years, and in order for life to have existed over this period the Sun must have been shining for at least this long, with a fairly constant energy output. The source of the Sun's energy could not have been chemical; if, for example, it was made out of coal which produces heat and light from chemical energy, then it would only have lasted for about 3000 years!

In the mid-1800s two physicists, Lord Kelvin (1824–1907) and Herman von Helmholtz (1821–94) put forward the idea that the huge weight of the Sun's outer layers should cause the Sun to gradually contract and, as it does so, the gases in its interior should become compressed. Now, whenever a gas is compressed its temperature increases – you may have seen this effect when a bicycle pump becomes warm after pumping up a tyre. Kelvin and Helmholtz argued that this gravitational contraction would cause the Sun's gases to become hot enough to radiate heat energy into space. This process, called the **Kelvin–Helmholtz contraction**, does in fact happen in the protostar and pre-main sequence phases of stellar formation. The loss of gravitational potential energy due to the contraction is converted into thermal radiation, making the gas glow weakly. However, the Kelvin–Helmholtz contraction cannot be the main source of stellar energy since, in the case of the Sun, calculations show that in order to produce the solar luminosity we see today, the Sun would have had to contract from a size larger than the Earth's orbit in only 25 million years.

Mass–energy equivalence

A clue to the source of stellar energy was discovered by Albert Einstein (1879–1955). In 1905, while developing his Special Theory of Relativity, Einstein showed that mass and energy are *equivalent*. If the energy of a body changes by an amount ΔE than its mass changes by an amount Δm given by the **mass–energy relation**:

$$\Delta E = \Delta m \times c^2$$

where c is the speed of light.

Now since c^2 is such a large number, a small amount of mass is equivalent to a large amount of energy. Thus the equivalent energy of 1 kg of matter is $1\ \text{kg} \times (3 \times 10^8\ \text{m s}^{-1})^2 = 9 \times 10^{16}$ joules, which is approximately the energy output of a 200 MW power station running for 14 years!

Earlier, we said that the source of stellar energy is nuclear. The nucleus forms some 99.98% of the atomic mass but occupies only 10^{-12} of its volume. How then, can this matter be converted into stellar energy?

The nuclear atom: fission and fusion

Nuclear binding energy

If we compare the mass of an atomic nucleus with the sum of the masses of its individual protons and neutrons, it is found that nuclei have a mass which is *less* than the total mass of their protons and neutrons. It is this **mass defect** that explains why the nucleus should exist at all, since for nuclei containing more than one proton, the electrostatic coulomb repulsion force of their like charges should cause the nucleus to fly apart. For nuclei to be stable, there exists a **strong nuclear force** between the nucleons that is short range, attractive and can overcome the coulomb repulsion of the protons (Figure 7.1).

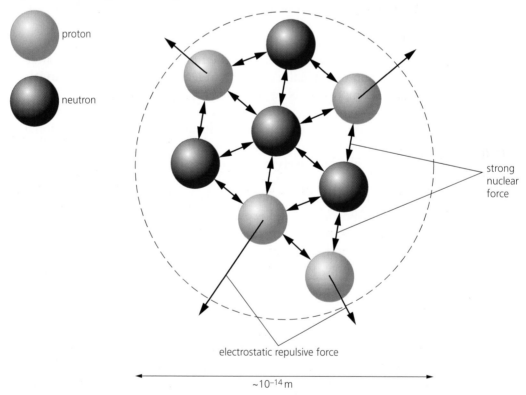

Figure 7.1 Forces within an atomic nucleus. The strong nuclear force overcomes the electrostatic repulsion of the protons and holds the nucleus together

Now suppose we assemble a nucleus of N neutrons and Z protons. There will be an *increase* in the electric potential energy due to the electrostatic forces between the protons trying to push the nucleus apart, but there is a *greater decrease* of potential energy due to the strong nuclear force acting between the nucleons, attracting them to one another. As a consequence, the nucleus has an overall *net decrease* in its potential energy. This decrease in potential energy is called the **nuclear binding energy** and the decrease per nucleon is called the **binding energy per nucleon**. This loss of energy is, by the mass–energy relation, equivalent to a loss of *mass*, hence the difference in the mass of a nucleus and the sum of its nucleons. It is the binding energy that holds the nucleus together, and its release is where the useful energy of the nucleus comes from (Figure 7.2).

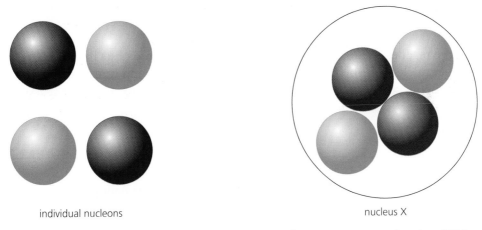

individual nucleons   nucleus X

Figure 7.2 Mass defect Δm = mass of N neutrons + mass of Z protons − mass of nucleus $^{N+Z}_{Z}X$. By the mass–energy relation, binding energy = $\Delta m \times c^2$

WORKED EXAMPLE 7.1

Helium $^4_2$He has 2 protons and 2 neutrons. The mass defect between an assembled helium nucleus and its individual nucleons is 0.0305 u, where u is the **unified atomic mass constant** (see Box 7.1). What is the binding energy of the helium atom?

$$\text{mass defect } \Delta m = 0.0305 \times 1.66 \times 10^{-29} \text{ kg}$$
$$\therefore \text{ change in energy } \Delta E = \Delta mc^2$$
$$= 5.06 \times 10^{-29} \text{ kg} \times (3.00 \times 10^8 \text{ m s}^{-1})^2 \text{ J}$$
$$= 4.6 \times 10^{-12} \text{ J}$$

We can represent the binding energy per nucleon as a graph which shows how it varies with atomic mass number A (Figure 7.3). Notice that the shape of the curve for low mass numbers indicates that the binding energy increases as the number of nucleons in the nucleus increases. There are also some nuclei such as helium ($^4_2$He), carbon ($^{12}_6$C) and oxygen ($^{16}_8$O) that lie above the trend of the curve and have binding energies greater than their neighbours.

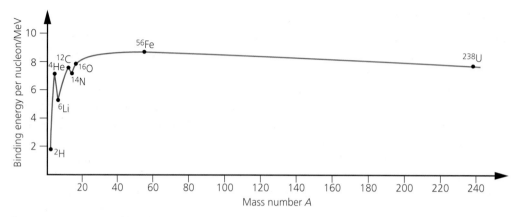

Figure 7.3 Variation of binding energy per nucleon with atomic mass number

The curve reaches a maximum at iron ($^{56}_{26}$Fe) which, because of its high binding energy per nucleon, results in the most stable nucleus. Beyond iron, the binding energy per nucleon falls slightly as A increases towards the more massive and sometimes unstable nuclei.

Box 7.1 The unified atomic mass constant

Since the mass–energy relation tells us that mass and energy are interchangeable, we introduce a measure of mass as an alternate unit of energy called the **unified atomic mass constant**, symbol u, which is defined to be $\frac{1}{12}$th of the mass of the carbon atom $^{12}_{6}$C.

One mole of carbon contains 6.02×10^{23} atoms of carbon and has a mass of 12 g. This means that:

$$\text{mass of 1 atom of carbon} = \frac{12}{6.02 \times 10^{23}} \text{ g} \quad \text{or} \quad \frac{12}{6.02 \times 10^{26}} \text{ kg}$$

But from our definition, this is equal to 12 u. Therefore:

$$1 \text{ u} = \frac{12}{12 \times 6.02 \times 10^{26}} \text{ kg} = 1.66 \times 10^{-27} \text{ kg}$$

Using the mass-energy relation we know that 1 kg of mass is equivalent to 9×10^{16} J so that:

$$1 \text{ u} = 1.66 \times 10^{-27} \times 9 \times 10^{16} = 1.49 \times 10^{-10} \text{ J}$$

Now 1 MeV = 1.6×10^{-13} J so

$$1 \text{ u} = \frac{1.49 \times 10^{-10}}{1.6 \times 10^{-13}} = 931 \text{ MeV (approx.)}$$

This conversion is used to change units of mass into MeV and since the mass of nucleons is very close to 1.66×10^{-27} kg, we see that the unified atomic mass constant is a very convenient quantity to use when calculating the energy released in nuclear reactions. Remember that:

$$1 \text{ u} = 1.66 \times 10^{-27} \text{ kg} = 931 \text{ MeV}$$

Fission and fusion

Two important processes that can release energy from the nucleus of an atom are **nuclear fission** and **nuclear fusion**. In both cases the useful energy is carried off by gamma rays and elementary particles.

In **nuclear fission**, a massive nucleus such as uranium splits into two to form two lighter nuclei of approximately equal mass. From Figure 7.3 we see that this happens on the *falling* part of the curve, so that mass is lost and binding energy is released when very heavy elements fission to nuclei of *smaller* mass number. Nuclear fission is responsible for the release of energy in nuclear reactors and atomic bombs.

In **nuclear fusion**, energy is released when two light nuclei are *fused* together to form a heavier nucleus. This happens on the *rising* part of Figure 7.3. Again mass is lost and binding energy released, and this is the principal source of energy in stars. (A hydrogen bomb also releases energy by nuclear fusion.) But since the nuclei of atoms carry positive charge due to their protons, how can they fuse together if they are repelled by the coulomb force acting between them? The answer is that fusion can happen if each nucleus has sufficient *kinetic energy* to enable them to overcome their mutual repulsion and be captured by the strong nuclear force.

In star formation, the kinetic energy needed for fusion comes from the conversion of gravitational energy into thermal energy by the Kelvin–Helmholtz contraction. Remember that a protostar forms when a cloud of interstellar hydrogen gas contracts under gravity. As it does so, the hydrogen atoms gain speed and therefore kinetic energy. Temperature is simply a measure of the average kinetic energy of the particles involved, so if protons in the hydrogen gas gain KE then the temperature of the cloud must also increase. In the case of stars like the Sun, fusion can occur when the temperature of the contracting cloud reaches about 8×10^6 K. It is because of the high temperatures needed to give the protons sufficient kinetic energy that these nuclear reactions are also known as **thermonuclear** fusion reactions (see Box 7.2).

Hydrogen burning

If stars produce energy by the fusion reaction, then the most likely nuclei involved are those that are most abundant in interstellar clouds, namely hydrogen nuclei. Also, hydrogen has the smallest nuclear electric charge – the single charge on a proton. Since like electric charges repel and the larger the nuclear charge the greater the repulsive force, nuclei with a small amount of charge are more likely to fuse together. We are led to the conclusion that:

It is the fusion of hydrogen nuclei by thermonuclear fusion reactions with a release of binding energy that is the primary source of energy generation in stars.

This is a very important process in the astrophysics of stars and is called **hydrogen burning**. Although nothing is 'burnt' in the ordinary sense of the word, what astrophysicists mean by this phrase is that hydrogen is converted to helium and the binding energy released is the explanation for a star's tremendous energy output.

A hydrogen nucleus consists of a single proton ($^1_1$H) whereas helium nuclei have two protons and two neutrons ($^4_2$He). Stars can make helium from hydrogen by fusing together *four* hydrogen nuclei:

$$4H \longrightarrow He + \text{energy released}$$

For this reaction to occur, two of the four H protons are converted into neutrons to create a single helium nucleus. How much energy is released in this process? We can work out how much mass is lost and thus the energy released:

$$\text{mass of 4H atoms} = 4 \times 1.008 \text{ u} = 4.032 \text{ u}$$
$$\text{mass of 1He atom} = 4.003 \text{ u}$$
$$\therefore \text{ mass loss } \Delta m = (4.032 - 4.003) \text{ u} = 0.029 \text{ u}$$

Using the mass–energy relation $\Delta m = \Delta E/c^2$ the equivalent energy ΔE is:

$$\Delta E = (0.029 \times 1.66 \times 10^{-27}) \text{ kg} \times (3 \times 10^8 \text{ m s}^{-1})^2$$
$$= 4.33 \times 10^{-12} \text{ J}$$

or, using electronvolts, since 1 u = 931 MeV,

$$\Delta E = 0.029 \times 931 = 27 \text{ MeV}$$

Box 7.2 Core temperature for hydrogen fusion

We can attempt to estimate the temperature needed for the fusion of hydrogen nuclei in the core of the Sun by equating the kinetic energy of a proton (hydrogen nucleus) at a temperature T in kelvin with the electrostatic potential energy of a helium nucleus containing two protons separated by a nuclear distance $r = R$, as follows.

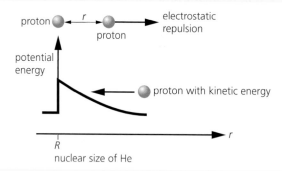

Figure 7.4 (a) Protons have to overcome their mutual electrostatic repulsion for fusion to occur

The average kinetic energy of a particle at a temperature T is given by KE $= \frac{3}{2}kT$ where k is Boltzmann's constant. The electrostatic potential energy of a proton a distance r from another proton is given by:

$$\frac{1}{4\pi\epsilon_0}\frac{e^2}{r}$$

where e is the electronic charge on the proton and ϵ_0 is the permittivity of free space. For fusion to occur, the kinetic energy of the proton must be at least equal to, or greater than, the electrostatic potential energy possessed by the helium nucleus. So we can write:

$$\tfrac{3}{2}kT \geqslant \frac{1}{4\pi\epsilon_0}\frac{e^2}{R}$$

$$\therefore T \geqslant \frac{1}{6k\pi\epsilon_0}\frac{e^2}{R}$$

Substituting the values: $R = 10^{-15}$ m (the approximate size of the helium nucleus, which is one of the smaller nuclei)

$$e = 1.6 \times 10^{-19}\,\text{C for the proton charge}$$
$$k = 1.38 \times 10^{-23}\,\text{J K}^{-1}$$
$$\epsilon_0 = 8.85 \times 10^{-12}\,\text{F m}^{-1}$$

we obtain:

$$T \geqslant \frac{(1.6 \times 10^{-19})^2}{6 \times 1.38 \times 10^{-23} \times \pi \times 8.85 \times 10^{-12} \times 10^{-15}}\,\text{K} \approx 10^{10}\,\text{K}$$

The actual temperature in the Sun's core, however, is much less than this – about 10^7 K. Why the discrepancy?

Figure 7.4b shows how the potential energy varies as a function of distance from the centre of helium nucleus.

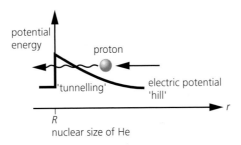

Figure 7.4 (b) Protons with lower kinetic energy may sometimes 'tunnel' through the potential barrier and be captured by the strong force

For large values of r the potential energy is due to the coulomb repulsion of the protons. At the surface of the nucleus ($r = R$), the potential energy falls drastically due to the action of the nuclear force and inside the nucleus ($0 < r < R$), it is constant. Quantum mechanics tells us that protons can 'tunnel' through this **potential barrier** without having to have enough energy to climb over the 'hill' and 'fall in' to the nucleus. This means that for fusion to occur, the minimum average kinetic energy is in reality much lower and so the temperature can be less.

Energy generation in the Sun

Let's now look more closely at how energy is generated in the Sun. We have seen that the fusion of four H atoms into a single He nucleus releases 4.33×10^{-12} J and this is quite a small amount of energy. However, look at it another way. About 0.7% of the total mass of the hydrogen we started with is 'lost' after the helium is created from the fusion reaction, so from 1 kg of H we produce 0.993 kg of He. Now 0.007 kg of mass when converted into energy via the mass–energy relation is $0.007 \times (9 \times 10^{16}) = 6.3 \times 10^{14}$ J and this is a *very* large amount of energy, comparable to burning 200 tonnes of coal.

We know that the luminosity of the Sun is 3.90×10^{26} W (J/s) and since the conversion of 1 kg of H into He yields 6.3×10^{14} J of energy, then approximately 3.90×10^{26} J/s $\div$ 6.3×10^{14} J $= 6.2 \times 10^{11}$ kg of H must be converted into He *every second*. This may seem like a huge consumption of hydrogen, but don't worry as the Sun has sufficient reserves of H for it to go on shining for at least another 5000 million years!

There are two principal nuclear reaction pathways in which hydrogen burning occurs in stars. These are the **proton–proton chain** and the **carbon–nitrogen–oxygen** or **CNO cycle**. In each of these reactions, four protons combine by nuclear fusion to form a single He nucleus with a small loss of mass as we have seen, which by the mass–energy relation is released as energy. The *temperature* in the interior of a star determines which one of these reactions takes place. For stars that have masses not exceeding that of the Sun, the temperature in the core of the star does not get higher than about 16×10^{6} K and hydrogen burning occurs via the proton–proton chain. In stars with masses greater than the Sun, the core temperatures exceed this value and hydrogen burning proceeds through the CNO cycle (see page 143).

The proton–proton chain

The proton–proton chain (p–p chain) proceeds in three steps. (At each step the energy released is expressed in MeV and is shown in brackets.)

Step 1 Two protons ($^{1}_{1}H$) fuse to form an isotope of hydrogen called deuterium ($^{2}_{1}H$), i.e.

$$^{1}_{1}H + ^{1}_{1}H \longrightarrow ^{2}_{1}H + ^{0}_{+1}e + \nu \qquad (1.19 \text{ MeV})$$

where $^{0}_{+1}e$ is an anti-electron called a **positron** (a particle that has the same mass as electron but a positive electric charge) and ν is an elementary particle which has zero charge and very low (possibly zero) mass, called a **neutrino**. The neutrino is a ghostly particle which only interacts occasionally with matter and escapes from the Sun in a few seconds, carrying away energy into space. Inside the Sun, the positron rapidly collides with a free electron and the two particles are totally annihilated with their mass being converted directly into energy in the form of gamma rays.

Step 2 A third proton then fuses with the deuterium nucleus to form a light isotope of helium ($^{3}_{2}He$) containing two protons and a single neutron. The energy released is carried away by a gamma ray (γ):

$$^{1}_{1}H + ^{2}_{1}H \longrightarrow ^{3}_{2}He + \gamma \qquad (5.49 \text{ MeV})$$

Step 3   In the last step, two $_2^3$He nuclei fuse together to produce an ordinary helium nucleus ($_2^4$He) plus two protons:

$$_2^3\text{He} + {}_2^3\text{He} \longrightarrow {}_2^4\text{He} + {}_1^1\text{H} + {}_1^1\text{H} \qquad (12.85\ \text{MeV})$$

It is important to remember that very high temperatures and densities are needed for hydrogen burning to occur. The temperature in the *core* of the Sun, that is in the inner quarter of its radius (see Figure 7.8), must be at least 8×10^6 K (see Box 7.2) with densities ranging from 1.6×10^5 kg m^{-3} in the centre to about 2×10^4 kg m^{-3} at the core perimeter.

Energy transport

How does the energy produced by hydrogen burning in the Sun's interior find its way to the surface of the Sun and escape as sunlight?

The Sun is in a state of **hydrostatic equilibrium**. This means that while the gravitational force due to the tremendous mass of its outer layers pushes inwards, trying to make it contract, as it does so, the internal gas pressure in the solar interior pushes outwards. The greater the compressive force, the higher the internal pressure. Hydrostatic equilibrium is reached when the gas pressure inside the Sun at every level can support the weight of the layers above (Figure 7.5).

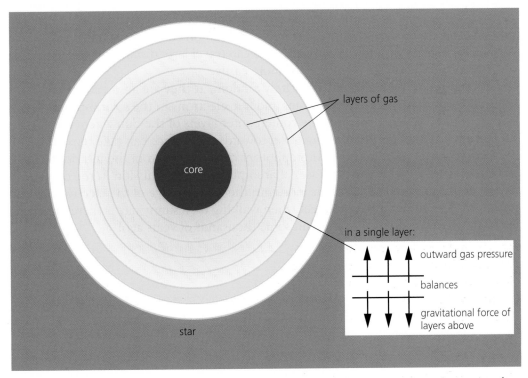

Figure 7.5 Hydrostatic equilibrium. The gravitational and pressure forces throughout the interior of the star are in balance, so it neither expands nor contracts

The Sun is also in a state of **thermal equilibrium**. The energy it radiates when shining is being replaced by energy from the interior at a steady rate so that it maintains a constant temperature. Remember that the Sun is a good approximation to a blackbody.

In general, heat energy can be transported by three processes: conduction, convection and radiation. **Conduction** is the process that occurs when, for example, one end of an iron bar is heated and the other end starts to get warm. In a gas, heat conduction is transmitted by the collisions of atoms or molecules, with those that have more kinetic energy transferring some to those with less. By using theoretical models, astrophysicists have found that the conditions inside the Sun do not particularly favour conduction as an efficient energy transport process, mainly because the solar material becomes less dense outside the core. In a main sequence star like the Sun, energy transport is by **convection** and **radiative diffusion**.

Convection is the process whereby heat is transferred from one place to another by movement of a fluid. Convection in the Sun occurs when hot gases rise towards its surface and cooler gases sink back down. As a result, circulation currents are set up in which heat energy is transferred to the outer layers of the Sun from its interior.

Recall that in the proton–proton chain, gamma photons are created that carry away energy. In **radiative diffusion**, these photons diffuse outwards from the hot core towards the surface. A property called the **opacity** affects how efficient the transfer of radiative energy is. Opacity is a measure of how easily a material absorbs photons. A gas with a high opacity means that it is difficult for radiative energy to flow through it; photons only travel short distances before they are absorbed so energy transfer is inefficient. In a gas with low opacity, photons can travel large distances before absorption and the energy is transferred more easily.

The motion of these photons is entirely random, as they are absorbed and re-emitted when they interact with atoms and free electrons in the solar interior. Their net motion, however, is towards the cooler, outer layers of the Sun where they escape into space. This photon migration towards the surface can take tens of thousands of years and in this fashion photons carry energy from the interior to the outside (Figure 7.6). By the processes of absorption and re-emission in the outer layers, emission from the surface is mostly in the visible range of the electromagnetic spectrum. The sunlight that you feel when walking on a sunny day is therefore due to photons that were created in the Sun some thousands of years ago!

Figure 7.6 Radiative diffusion. Photons migrate from the core of the Sun and, because of its opacity, follow a random path as they travel to the surface, taking thousands of years to do so

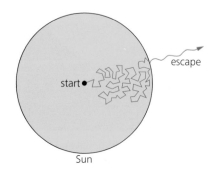

start

escape

Sun

Stellar models

The physical concepts of hydrostatic equilibrium, thermal equilibrium and energy transport can be combined together to form a set of mathematical equations called the **equations of stellar structure** which enable astrophysicists to construct a model of a star which describes the physical conditions occurring inside it.

Using supercomputers, astrophysicists can solve these equations to calculate the necessary conditions of pressure, temperature and density inside a star which must be satisfied if the star is to remain in a stable form. The astrophysicist begins with observational data about the star, such as its luminosity, surface temperature and what can be deduced about the gas pressure and density changes with increasing depth all the way down to the centre of the star. The results constitute a **stellar model** of the star which is often presented in a table or as a series of graphs.

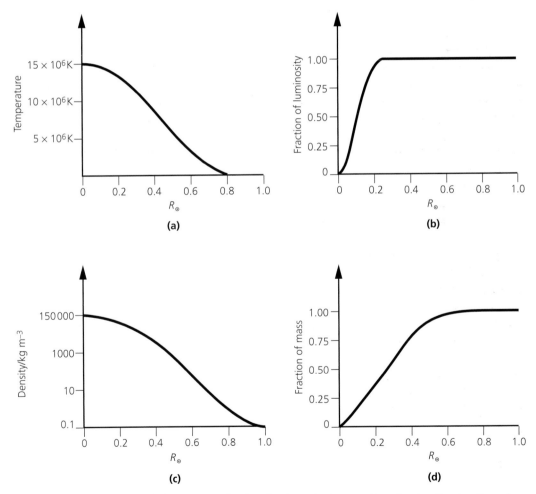

Figure 7.7 Stellar model of the Sun: graphs showing how the temperature, luminosity, density and mass vary with distance from the Sun's centre (in solar radii)

Table 7.1 Stellar model of the Sun in the form of a table

Fraction of solar radius	Temperature/ 10^6 K	Fraction of central pressure	Density/ kg/m³	Fraction of mass	Fraction of luminosity
0.0	15.5	1.00	150 000	0.00	0.00
0.1	13.0	0.46	90 000	0.07	0.42
0.2	9.5	0.15	40 000	0.35	0.94
0.3	6.7	0.04	13 000	0.64	1.00
0.4	4.8	0.007	4 000	0.85	1.00
0.5	3.4	0.001	1 000	0.94	1.00
0.6	2.2	0.003	400	0.98	1.00
0.7	1.2	4×10^{-5}	80	0.99	1.00
0.8	0.7	5×10^{-6}	20	1.00	1.00
0.9	0.3	3×10^{-7}	2	1.00	1.00
1.0	0.006	4×10^{-13}	3×10^{-4}	1.00	1.00

(radiative zone: 0.0–0.7; convective zone: 0.8–1.0)

Figures 7.7a–d and Table 7.1 show a stellar model of the Sun. In Figure 7.7b, notice that the fraction of luminosity with radius increases to 100% at a radius of $0.25R_\odot$ from the Sun's centre and from this we can deduce that the Sun's energy production must take place in a volume within a quarter of its radius. Also, from Figure 7.7d, at 0.8 solar radii the fraction of the Sun's mass is nearly 100%. As the Sun's radius is about 700 000 km, this implies that the outer 140 000 km consists of very little solar matter, which also accounts for the low density and pressure in this region (see Table 7.1).

In these outer layers of the Sun, convection is the dominant transport process and the region from about $0.8R_\odot$ is called the **convective zone**. From the centre to $0.8R_\odot$ (560 000 km), radiative energy transport dominates, and this region is called the **radiative zone**.

It is the equations of stellar structure that enable us to illustrate the interior of the Sun as in Figure 7.8.

The solar neutrino problem

A puzzling feature of the Sun's stellar model is the **solar neutrino problem**. Remember that neutrinos are the ghostly particles produced in the p–p chain and escape from the Sun almost immediately. Neutrinos are very difficult to detect because they interact very weakly with matter. Approximately 10^{38} neutrinos are generated in the Sun every second, and even the Earth with its large mass is virtually transparent to them. Astrophysicists are interested in detecting neutrinos because they provide a window into the thermonuclear reactions occurring in the Sun's core. Various neutrino detectors have been constructed; one is a big tank filled with tetrachloroethylene (C_2Cl_4), commonly used as dry-cleaning fluid, located deep underground in a gold mine in Homestake, South Dakota, USA (Figure 7.9).

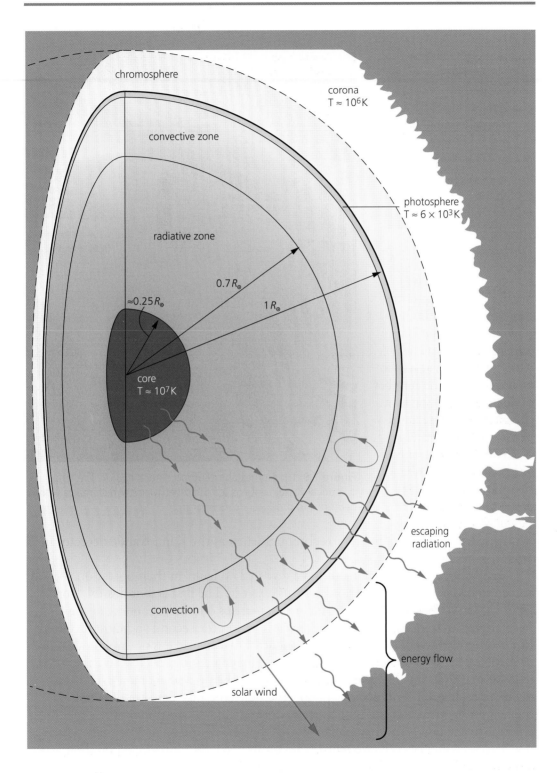

Figure 7.8 The interior of the Sun

About 10^{14} neutrinos per square metre arrive at the Earth from the Sun each second and most of these pass straight through, but occasionally one neutrino (ν) will react with the nucleus of a chlorine atom in the Homestake detector, producing a radioactive isotope of argon ($^{37}_{18}\text{Ar}$) and an electron:

$$^{37}_{17}\text{Cl} + \nu \longrightarrow {}^{37}_{18}\text{Ar} + \text{e}^-$$

Using a very sensitive method, the argon atoms can be flushed from the tank and their number measured. The rate at which the argon atoms are created tells us the number of neutrinos arriving from the Sun per square metre per second. The strange observation is that only *one-third* of the number of solar neutrinos as predicted by the stellar model are found. Other

Figure 7.9 The Homestake neutrino detector, USA. The long, cylindrical vessel contains tetrachloroethylene, which reacts with neutrinos to produce atoms of argon

neutrino detectors using different techniques (such as the Kamiokande detector in Japan) so far confirm this.

Careful checks of the equipment have revealed no faults in their operation and we are led to two possible explanations for this mystery. The first is that possibly our construction of the stellar model is incorrect; perhaps our understanding of the nuclear physics is incomplete in some way. However, the second and most favoured explanation is that our knowledge of the neutrino as a particle is not well understood.

Particle physicists who study elementary and fundamental particles have theorised that the neutrino can change from one kind to another in a phenomenon called a **neutrino oscillation**. This could explain why the Homestake detector does not see them, if for two-thirds of the time they are not the kind of neutrino that react with the chlorine in the tank. Another explanation is that the neutrinos may actually possess a non-zero mass and be converted into other forms of energy, or particles, on their way to Earth. If this is so, then important cosmological ideas such as the total density of matter and energy in the universe would need to be reconsidered (see Chapter 10).

The CNO cycle

While the Sun burns hydrogen by the proton–proton chain, this is not the case for more massive stars whose core temperature exceeds 16 million K. There hydrogen burning is achieved by the **CNO cycle**. In the CNO cycle the presence of *carbon* plays an important role in converting hydrogen into helium. The cycle has six steps. (As before, the energy released at each step is shown in brackets.)

Step 1 A carbon atom and a proton fuse to form an isotope of nitrogen and a γ ray:

$$^{12}_{6}C + {}^{1}_{1}H \longrightarrow {}^{13}_{7}N + \gamma \quad (1.95 \text{ MeV})$$

Step 2 The isotope of nitrogen decays to an isotope of carbon, emitting a positron (e^+) and a neutrino in the process:

$$^{13}_{7}N \longrightarrow {}^{13}_{6}C + {}^{0}_{+1}e + \nu \quad (2.22 \text{ MeV})$$

Step 3 The isotope of carbon fuses with another proton to form stable nitrogen and a γ ray:

$$^{13}_{6}C + {}^{1}_{1}H \longrightarrow {}^{14}_{7}N + \gamma \quad (7.54 \text{ MeV})$$

Step 4 Stable nitrogen fuses with a third proton to form an isotope of oxygen and γ ray:

$$^{14}_{7}N + {}^{1}_{1}H \longrightarrow {}^{15}_{8}O + \gamma \quad (7.35 \text{ MeV})$$

Step 5 The isotope of oxygen decays to form an isotope of nitrogen, a positron and a neutrino:

$$^{15}_{8}O \longrightarrow {}^{15}_{7}N + {}^{0}_{+1}e + \nu \quad (2.71 \text{ MeV})$$

Step 6 Finally, a helium nucleus is created when the isotope of nitrogen fuses with a fourth proton and the carbon-12 is restored:

$$^{15}_{7}N + {}^{1}_{1}H \longrightarrow {}^{12}_{6}C + {}^{4}_{2}He \quad (4.96 \text{ MeV})$$

Notice that, as in the p–p chain, the CNO cycle takes four hydrogen nuclei (protons) and converts them into a single helium together with positrons, neutrinos and some high-energy gamma rays. The $^{12}_{6}C$ nucleus acts as a catalyst for the reaction and while it is consumed in Step 1, it is replaced in Step 6, so that in the CNO reaction chain as a whole, carbon is not used up.

Post-hydrogen burning

So far we have considered how stars, after collapsing out of giant molecular clouds, become hot enough for thermonuclear fusion reactions to begin. As they start to shine they convert hydrogen into helium but what happens when the hydrogen runs out? The answer to this question is part of a wider one which addresses how stars evolve over their lifetimes and also explains the origin of many of the chemical elements in the periodic table.

In Chapter 8, we will see how other nuclear fusion reactions come into play and how heavier elements are 'cooked' inside stellar interiors, which in turn contribute to the abundances of the elements that are found in the universe. We will see that if it were not for the nuclear reactions that occur deep inside stars, then none of the heavier elements (nor you and I!) would exist in the universe today.

Summary

◆ Neither chemical nor gravitational energy is sufficient to account for the enormous energy generated by stars. Einstein showed that matter and energy are equivalent by means of the **mass–energy relation** $\Delta E = \Delta mc^2$. The large numerical value of c^2 means that a small amount of mass is equivalent to a large amount of energy. The source of stellar energy is nuclear.

◆ The nucleus of an atom is held together by **nuclear binding energy**. Energy may be released from an atom by **nuclear fission** or **fusion**. In main sequence stars **hydrogen burning**, where four hydrogen nuclei undergo fusion to form helium, is the primary energy production process. The resulting mass loss is converted into energy by the mass–energy relation.

◆ For nuclear fusion to occur, the kinetic energy of protons must be equal to or greater than the electrostatic potential energy at nuclear separations caused by their mutual repulsion. The average kinetic energy of the protons is proportional to the temperature and so for nuclei to fuse together very high temperatures are needed.

◆ Hydrogen burning is responsible for energy generation in the Sun. It occurs in the **core**, i.e. the inner quarter of the Sun's radius. Energy is transported to the outer layers of the Sun (and similar main sequence stars) by **convection** and **radiative diffusion**. The Sun has sufficient reserves of hydrogen to go on shining for thousands of millions of years.

◆ Hydrogen burning can proceed in two ways. The three-step **proton–proton (p–p) chain** occurs in low mass stars whose core temperatures do not exceed 16 million K and six-step **carbon–nitrogen–oxygen (CNO) cycle**, in which carbon acts as a catalyst in converting hydrogen into helium, occurs in more massive stars.

◆ A star in a stable state is in **hydrostatic equilibrium**. This means that at all points, its outward internal gas pressure is balanced by the inward force due to its mass, so that it neither expands nor contracts.

◆ **Stellar models** based on observational data can be constructed which give information about the pressure, temperature and density at different levels inside a star. These can be compared with experimental measurements. A significant problem arising from the stellar model of the Sun is the **solar neutrino problem**.

◆ Hydrogen burning will not last for ever, but is one step in the manufacture of the chemical elements. It is succeeded by different nuclear reactions as the hydrogen is used up and the star proceeds to its next stage of stellar evolution.

Questions

1 What is the difference between nuclear fission and nuclear fusion? Which process is responsible for energy generation in stars?

2 In 1945 an American B29 bomber, the *Enola Gay*, dropped an atomic bomb on Hiroshima which had an energy equivalent of 20 000 tons of TNT or 8.4×10^{19} J. What mass in kg was converted into energy in the detonation?

3 Explain what is meant by the term *mass defect*. What is the relationship between the mass defect and the *binding energy* of a nucleus?

4 Experiments on Earth have been carried out to try to produce controlled energy from nuclear fusion in a *fusion reactor* using the reaction:

$$\ _1^2H + \ _1^2H \longrightarrow \ _2^3He + \ _0^1n$$

where $_1^2H$ is an isotope of hydrogen called a *deuteron* and $_2^3He$ is an isotope of helium. Using the mass–energy relation, calculate the energy released in this reaction in MeV. (Mass of deuteron = 2.015 u; mass of $_2^3He$ = 3.017 u; mass of $_0^1n$ = 1.009 u)

5 What are the two main processes by which energy is transported from the core of the Sun to its surface?

6 In the second step of the p–p cycle, a gamma ray photon of 5.49 MeV is produced that carries away radiative energy. What is its wavelength?

7 What is meant by *hydrostatic equilibrium*?

8 a What is a *stellar model*? What information do stellar models give to help astrophysicists understand the interior of stars?
   b What is the solar neutrino problem?

9 Explain what is meant by *hydrogen burning*. What happens in the p–p chain and CNO cycle? For nuclear fusion of hydrogen very high temperatures are needed. Explain why higher temperatures are required in the core for hydrogen fusion to proceed by means of the CNO cycle compared with when it proceeds via the p–p chain.
   If in the Sun 3×10^{14} moles of hydrogen are converted into helium nuclei by the p–p chain every second, estimate the number of joules per second radiated by the Sun. (Mass of H nucleus = 1.0078 u; mass of helium nucleus = 4.0026 u)

10 The Sun shines by converting matter into energy. This implies that it now 'weighs' slightly less than when you started reading this chapter. Its luminosity is 4×10^{26} J s^{-1}. How much mass is lost per second by the Sun as it shines?
   If the mass of the Sun is 2×10^{30} kg, assuming that it is all hydrogen and carries on shining by hydrogen burning, for how many years can it shine before all its mass is consumed?

11 Outline the stages by which a large mass of cold hydrogen becomes a main sequence star. Include in your explanation an account of the energy transfer processes within the star but *not* the details of the pp chain.

London Physics, June 1999, part

The lives of stars and stellar nucleosynthesis

In Chapter 5, we saw how the Hertzsprung–Russell Diagram (HRD) showed that there are different types of star. Most stars lie along a straight line called the main sequence while the smaller white dwarfs and larger red giants congregate together in groups. In this chapter we will see how the HRD has much to tell us about the lives of stars: how their mass determines their position on the diagram and how they might evolve. We will also see how stars make many of the chemical elements found in the periodic table.

Stellar evolution and the Hertzsprung–Russell Diagram

So far, we have seen how stars form out of material in the interstellar medium, initially as protostars, then begin to shine by thermonuclear reactions and are held together by hydrostatic equilibrium.

During their lifetimes, stars evolve in a way that can be related to their position on the Hertzsprung–Russell Diagram (Figure 5.5). There are four main evolutionary stages in the life of a star: **protostar**, **pre-main sequence**, **main sequence** and **post-main sequence**. Stellar evolution is concerned with the slow changes in a star's size, luminosity and surface temperature with time, and the relationship between these and its chemical composition.

Protostars and pre-main sequence stars

In Chapter 6, we discussed in depth the processes by which stars form and begin to shine. To summarise: interstellar clouds contract under gravity and gravitational potential energy is converted into thermal and radiative energy. Eventually the core of the cloud becomes hot enough for nuclear fusion reactions to begin. While the star is still in free-fall collapse it is called a protostar. From this stage until nuclear ignition it is called a pre-main sequence star; internal pressure retards the collapse and it becomes hot enough to be faintly luminous.

There are exceptions to the progression, however. If a protostar reaches a mass less than about $0.08M_\odot$, then the gravitational contraction will be insufficient for nuclear reactions to start and the star will never reach the main sequence. Such an object is called a **brown dwarf** and is more like an enormous planet than a star. The only energy source for brown dwarfs is gravitational potential energy and as a consequence they are cool and have low luminosities.

Stellar models show that an object is classified as a planet if its mass is less than $0.002M_\odot$, a brown dwarf if its mass is between $0.002M_\odot$ and $0.08M_\odot$; and a star capable of hydrogen burning if the initial mass exceeds $0.08M_\odot$. It is estimated that at least 500 brown dwarfs may exist in a volume 20 pc in diameter within the disk of our own galaxy.

Evolution onto the main sequence

The path describing the progress of a star on the HRD is called its **evolutionary track**. A star in the pre-main sequence stage shrinks as the Kelvin–Helmholtz contraction causes it to glow. Its low surface temperature and luminosity means that an evolutionary track places it in the lower right-hand corner of the HRD (point A on Figure 8.1) As the star continues to contract, its core temperature increases to the point where nuclear reactions commence and supply the energy to make it shine.

Hydrostatic equilibrium is now established, and the star begins its useful working life. It is now called a **zero-age main sequence (ZAMS)** star (point B on Figure 8.1). At this point, it starts the longest stage in its career, burning hydrogen into helium over many millions of years. This era is known as the **main-sequence phase**.

As the hydrogen in the core is used up, hydrostatic equilibrium causes the interior temperature to gradually increase as well as the density. This ensures that the rate of nuclear reactions in the core are maintained. The star contracts slightly, increasing its luminosity due to a greater flow of energy to its surface. On the HRD the evolutionary track moves along the main sequence until it reaches point C, where it starts to veer over to the right and the star is now ready to enter the next stage of its evolution.

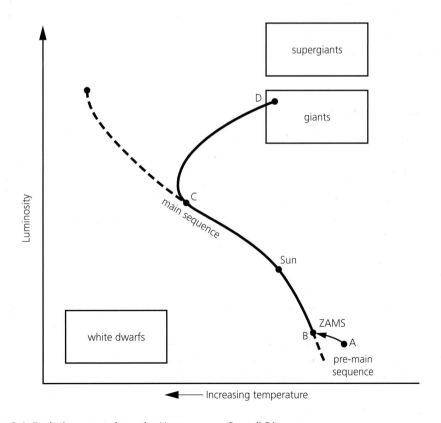

Figure 8.1 Evolutionary track on the Hertzsprung–Russell Diagram

Lifetime on the main sequence

How long the star spends on the main sequence depends on its *mass*. The Sun has a main sequence lifetime of roughly 10^{10} years (see Box 8.1). We can derive a simple expression for the lifetime of a star by using the mass–energy relation.

Box 8.1 The lifetime of the Sun

How long will the Sun shine due to hydrogen burning? A way to estimate this is to remember that the Sun converts hydrogen into helium by the proton–proton chain and the fusion of four H nuclei results in the production of one He nucleus. In Chapter 7, we saw that this fusion reaction releases 4.33×10^{-12} J of energy. We also know that the luminosity of the Sun is 4×10^{26} J s^{-1}.

If we consider the Sun as a ball of gas of average density 1400 kg m^{-3} consisting of mainly ionised hydrogen, the number N_p of particles in the Sun is therefore the mass of the Sun divided by the average mass of each proton (the mass of the electrons is negligible on this scale).

$$N_p = \frac{2 \times 10^{30}}{1.67 \times 10^{-27}} = 1.2 \times 10^{57}$$

and so the total energy released by fusion of all the H into He will be:

$$\frac{1.2 \times 10^{57} \times 4.33 \times 10^{-12} \text{ J}}{4} = 1.3 \times 10^{45} \text{ J}$$

(remembering to divide by 4 since it takes four H nuclei to make one He nucleus by the p–p chain).

The lifetime of the Sun shining by hydrogen burning is therefore the total energy released by fusion divided by the Sun's luminosity, so that:

$$T(\text{hydrogen burning}) = \frac{1.3 \times 10^{45} \text{ J}}{4 \times 10^{26} \text{ J s}^{-1}} = 10^{11} \text{ years}$$

In fact stellar models show that the real value is a factor of 10 smaller. This is because only 10% of the hydrogen is used up before the Sun proceeds to its next stage of evolution; the main sequence lifetime of the Sun is taken as 10^{10} years.

The energy released by a star during its hydrogen-burning phase will be its luminosity multiplied by the time over which hydrogen burning occurs. Let the change in energy by the mass–energy relation be $\Delta E = fMc^2$, where fM is the fraction of the star's equivalent mass. Then since $\Delta E = LT_{ms}$ where L is the luminosity and T_{ms} the hydrogen-burning period (on the main sequence), we can write:

$$T_{ms} = \frac{fMc^2}{L} \qquad \text{or} \qquad T_{ms} \propto \frac{M}{L}$$

So the lifetime of a star on the main sequence is proportional to its mass divided by its luminosity. For the Sun $T_{ms} = 10^{10}$ years, so that in general T_{ms} for a star can be written as:

$$T_{ms} = 10^{10} \frac{M_{star}}{M_{\odot}} \times \frac{L_{\odot}}{L_{star}} \text{ years}$$

In Chapter 5 we discussed the mass–luminosity relation represented by a graph of the luminosity L of a star plotted against its mass M (Figure 5.6). A relationship that approximately fits the points on this graph is of the form:

$$L \propto M^{3.5}$$

Therefore we can write:

$$T_{ms} \propto \frac{M}{M^{3.5}} \propto \frac{1}{M^{2.5}}$$

and this implies that more massive stars spend a shorter time on the main sequence than less massive ones.

   The more mass a star has, the more nuclear fuel is available to produce energy by hydrogen burning but, by the mass–luminosity relation, it radiates a greater number of joules per second. It is important to remember that *the higher the mass of a star, the shorter the lifetime*. Although a massive star has more nuclear fuel, it burns it at a faster rate since its luminosity is much higher.

Evolution post-main sequence

While on the main sequence, a star uses up less than 10% of its hydrogen reserves and so does not appreciably change its mass. However, in its core, where the nuclear reactions occur, the composition changes as hydrogen is converted into helium, forcing the star to radically alter its structure and appearance.

   When all the hydrogen in the core is exhausted, the thermonuclear reactions stop and gravity compresses the core to a smaller size. However, surrounding the core (which is now mainly helium) is a shell of hydrogen which is heated by the gravitational contraction, and hydrogen burning continues in the shell. This produces an outward pressure which prevents the star from collapsing and makes it expand to several hundred times its former size. It has now become a **red giant** with a lower surface temperature and higher luminosity (point D on Figure 8.1).

   Red giant stars are like very tenuous gas balls. Stellar models show that their density is very low, about 10^{-6} kg m^{-3}, and their surface temperatures are in the range 3000–4000 K. They have radii between $10R_{\odot}$ and $100R_{\odot}$ and are usually of spectral class M. Most of their mass is concentrated in the core which may only be the size of a few Earth diameters. Some examples of red giants that can be seen with the naked eye are *Arcturus* in the constellation *Bootes*, and *Aldebaran* in *Taurus*. A few stars are both more luminous and larger than typical red giants and are called **supergiants**. These have radii up to $1000R_{\odot}$ and appear in the top right-hand corner of the HRD. Examples of supergiants are *Betelgeuse* in *Orion*, and *Antares* in *Scorpius*.

Helium burning

For a period of several hundred million years it is only hydrogen burning in the shell that keeps that star shining while in the core the temperature continues to increase due to gravitational contraction. Inside the core the helium is highly ionised with helium nuclei and electrons freely moving about, and gravity has compressed the core to such a high density that the electrons become **degenerate** (degeneracy is discussed in detail in Chapter 9). This means that the core resists any further gravitational contraction and hydrostatic equilibrium is maintained, even though there are no thermonuclear fusion reactions taking place.

Eventually, at about 10^8 K, the core is hot enough for the helium nuclei to have enough thermal energy to overcome their mutual repulsion and fuse together. A new nuclear burning process called **helium burning** starts up, involving a chain of nuclear reactions called the **triple alpha process**. This occurs in two steps and is written as:

Step 1 $\quad {}^4_2\text{He} + {}^4_2\text{He} \longrightarrow {}^8_4\text{Be} + \gamma$

Step 2 $\quad {}^8_4\text{Be} + {}^4_2\text{He} \longrightarrow {}^{12}_6\text{C} + \gamma$

Two helium nuclei fuse together to form a beryllium nucleus and a gamma ray (Step 1). The beryllium nucleus immediately collides with a third helium nucleus to form ${}^{12}_6\text{C}$ and a gamma ray (Step 2). The triple alpha process is so-called since *three* helium nuclei are involved, forming a carbon nucleus (and helium nuclei are otherwise alpha particles). Some of the carbon created in the triple alpha process can fuse with a stray helium nucleus to produce oxygen by the reaction:

$$ {}^{12}_6\text{C} + {}^4_2\text{He} \longrightarrow {}^{16}_8\text{O} + \gamma $$

so that the 'ash' of helium burning can contain both carbon and oxygen. Stellar models indicate that a red giant burns helium in its core for approximately 5 to 20% of the time that it lived as a main sequence star burning hydrogen.

Stellar masses and nucleosynthesis

Low mass stars

The way in which helium burning starts in the core of a red giant depends on the star's initial mass. For *high* mass stars, in excess of $3M_\odot$, helium burning begins gradually, spreading throughout the core. However in *low* mass stars, less than $3M_\odot$, helium burning starts suddenly, in what astrophysicists call the **helium flash**, with a rapid rate of increase of the triple alpha process.

In the case of a low mass star, the helium flash initiates the steady conversion of helium into carbon in the core, with hydrogen burning in a shell surrounding it. This phase lasts for less than 10^8 years, leaving carbon as the leftover 'ash'. There is insufficient mass in low mass stars for core temperatures to increase by gravitational contraction to the point where the fusion of carbon is possible and so thermonuclear reactions in the core stop. For a while longer helium burning goes on in a shell around the core (and hydrogen in a shell around that) but eventually these cease as

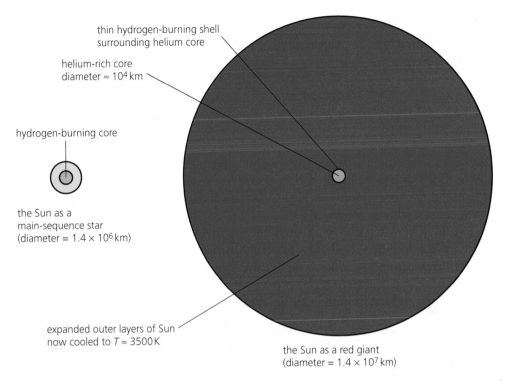

Figure 8.2 The interior of an old low mass star. Near the end of its life a low mass star becomes a red giant. Its internal structure consists of a helium-rich core with concentric shells of helium and hydrogen nuclear burning, all contained in a volume about the same size as the Earth

well. The outer envelope of the star is ejected into space in the form of a spherical shell of cooler and thinner matter called a **planetary nebula**. We will say more about the final states of stars in Chapter 9.

High mass stars

For high mass stars the situation is somewhat different. Stars of masses between 3 and $50M_\odot$ follow more or less the same pre-main sequence and post-main sequence phases. As their mass is larger, hydrogen burning takes place via the CNO cycle and the time spent on the main sequence is shorter. An important variation is that more massive stars reach *higher temperatures* in their cores and this has implications for the manufacture, or **nucleosynthesis**, of the heavier elements in the periodic table.

At ZAMS, the proton–proton chain initiates nuclear reactions in the core. However, as the core temperature increases, the CNO cycle becomes the dominant mechanism for hydrogen burning and the luminosity of the star increases. Although massive stars have more hydrogen fuel to consume, their luminosities are much higher and hydrogen burning proceeds at a faster rate. This explains why massive stars spend a shorter time on the main sequence despite having a greater initial store of hydrogen. Table 8.1 shows a comparison between stellar masses and their main sequence lifetimes.

Table 8.1 Characteristics of stars of different masses

Mass/$M_\odot$	Spectral type	Luminosity/$L_\odot$	Surface temperature/K	Main sequence lifetime/10^6 years
25–60	O	80 000–500 000	>35 000	3–1
4–20	B	800–2000	11 000–35 000	500–15
0.5–4	A, F, G, K	0.03–60	4000–11 000	200 000–500

Notice how dramatically the main sequence lifetime decreases with mass. While hot, massive stars shine more brightly they spend less time doing so. Our own Sun, while not being in the super-luminous league, is nonetheless a steady old plodder on the main sequence (thank goodness!).

When high mass stars have finished burning hydrogen in their cores, the process continues as before in a shell of hydrogen within their outer layers. The star expands into a red giant and the surface temperature drops. The core contracts and heats up to the point where the triple alpha process occurs, but in this case there is no helium flash triggering sudden helium burning, and the triple alpha process proceeds gradually.

Post-helium burning

After helium burning ceases in high mass stars, gravitational contraction causes the star's core temperature to increase to the point where **carbon burning** begins. In this process neon, helium, magnesium and oxygen can be produced by the following reactions:

$$^{12}_{6}C + {}^{12}_{6}C \longrightarrow {}^{20}_{10}Ne + {}^{4}_{2}He$$

$$^{12}_{6}C + {}^{12}_{6}C \longrightarrow {}^{24}_{12}Mg + \gamma$$

$$^{12}_{6}C + {}^{12}_{6}C \longrightarrow {}^{16}_{8}O + 2\,{}^{4}_{2}He$$

Remember that the higher the temperature in the centre of the star, the greater the kinetic energy available for nuclei to overcome their mutual electrostatic repulsion and fuse together. Carbon burning is possible for stars larger than $3M_\odot$; for stars with a main sequence mass of $9M_\odot$, the central temperatures become great enough for **neon burning** which produces oxygen, magnesium and helium by the nuclear reactions:

$$^{20}_{10}Ne + \gamma \longrightarrow {}^{16}_{8}O + {}^{4}_{2}He$$

$$^{20}_{10}Ne + {}^{4}_{2}He \longrightarrow {}^{24}_{12}Mg + \gamma$$

If the core temperature rises to about 1.5×10^9 K, oxygen nuclei can fuse and **oxygen burning** begins by the reaction:

$$^{16}_{8}O + {}^{16}_{8}O \longrightarrow {}^{32}_{16}S + \gamma$$

The principal product of oxygen burning is sulphur, although two isotopes of sulphur are produced as well as heavier nuclei such as silicon and phosphorus by alternative nuclear reaction pathways. These are shown below:

$${}^{16}_{8}\text{O} + {}^{16}_{8}\text{O} \text{ can fuse to form} \begin{cases} {}^{28}_{14}\text{Si} + {}^{4}_{2}\text{He} + \gamma \\ {}^{31}_{7}\text{P} + {}^{1}_{1}\text{H} + \gamma \\ {}^{31}_{8}\text{S} + \text{n} + \gamma \\ {}^{24}_{12} + {}^{4}_{2}\text{He} + {}^{4}_{2}\text{He} \\ {}^{30}_{14}\text{Si} + {}^{1}_{1}\text{H} + {}^{1}_{1}\text{H} + \gamma \end{cases}$$

Notice that one of these oxygen-burning pathways also produces neutrons. Since neutrons have no electric charge they can easily collide with other nuclei and be captured to form new isotopes. This process of **neutron capture** can produce many other elements not produced by fusion reactions (see Box 8.2).

Finally, in massive stars, the central temperatures reach high enough values (greater than 3×10^9 K) so that **silicon burning** can begin. The fusion of two silicon nuclei can give rise to many different nuclear reactions and the final nucleus produced is that of stable iron (${}^{56}_{26}\text{Fe}$). Iron has the highest binding energy per nucleon (see Figure 7.3, page 132), and no more energy can be released by fusing iron nuclei together. The core of the star becomes iron-rich although shell burning is still going on in the star's outer layers.

Each stage of thermonuclear burning is progressively shorter. In a star of mass $25M_\odot$, stellar models suggest that carbon burning takes 600 years, neon burning 1 year and oxygen burning only 6 months. In silicon burning the nuclear reaction rate is so high that the silicon in the core is converted to iron in about 1 day!

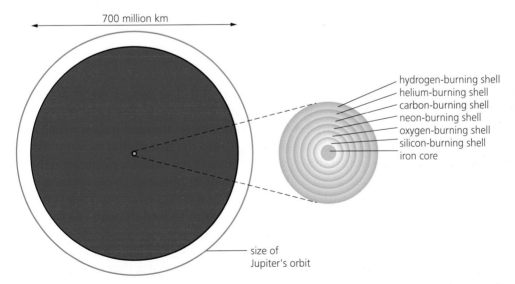

Figure 8.3 The interior of an old high mass star. The star's energy comes from a series of concentric burning shells synthesising different elements which are all within a volume about the same size as the Earth

Box 8.2 The manufacture of heavy elements – the r- and s-processes

Stellar cores can burn in such a way as to produce elements heavier than iron. Starting with hydrogen burning, fusion reactions at higher and higher core temperatures produce many elements up to iron, with the 'ashes' of preceding reactions provide fuel for successive ones. At iron, however, fusion stops because to fuse iron to heavier elements requires an energy *input*. The synthesis of elements beyond iron takes place in supernovas and red giants and involves the capture of *neutrons*.

Many reactions in stellar cores (such as the fusion of oxygen) produce free neutrons and because these have zero charge, they can easily penetrate the nucleus and be captured by the strong force. In doing so neutrons are added to iron and other elements, producing neutron-rich nuclei. Of the elements produced and their associated isotopes, many are unstable and decay by **beta decay**; a neutron changes into a proton emitting an electron and neutrino by the reaction:

$$n \longrightarrow p + \beta + \nu$$

The rate of β decay depends on the *rate of neutron capture*. If a nucleus captures neutrons *faster* than it can change them into protons by β decay, then neutron-rich nuclei are formed and this process is called a rapid or **r-process**. Alternatively, if a nucleus captures neutrons *slowly* then the nucleus can undergo β decay *before* capturing another neutron and proton-rich nuclei are formed. This process is called the slow or **s-process**.

The r- and s-processes are responsible for the synthesis of most of the elements heavier than iron. In the case of red giants, the s-process operates during helium burning; elements heavier than iron but lighter than lead are manufactured and blown off into the interstellar medium by the red giant's stellar wind. In supernovas, due to their short-life nature, nucleosynthesis occurs mainly by the r-process and elements heavier than lead, such as uranium and thorium, are made and then expelled into the ISM with explosive violence!

Mass loss from stars

A red giant has a very low surface gravity because of its low density and there is a steady stream of matter flowing out from its surface into space. For a low mass star, bursts of triple alpha reactions in the star's interior cause **thermal pulses** which rip off the star's outer gaseous envelope and blow it into the interstellar medium. As we mentioned earlier, the result is the expanding shell of gas called a planetary nebula and a low mass star can lose more than half its mass in this way. (Note: planetary nebulas have nothing to do with planets. The term was coined in the 18th century because these glowing objects resembled planets when viewed through the small telescopes available at the time.)

Figure 8.4 shows a planetary nebula called the *Cat's Eye Nebula* in the constellation of *Draco* about 3000 ly away. The hot central star that ejected the expanding shell can be seen in the centre. The central star radiates strongly at ultraviolet wavelengths, causing the gases in the shell to fluoresce which gives the nebula a glowing appearance. The planetary nebula phase of a star is quite short.

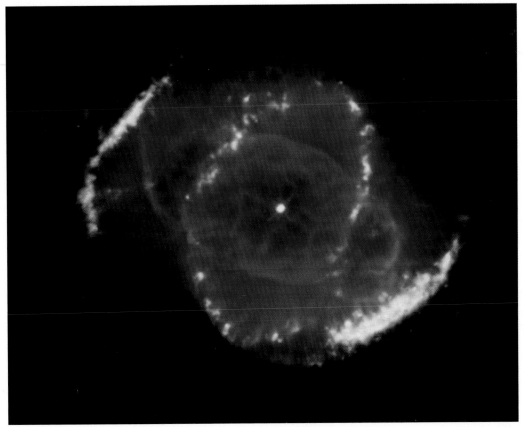

Figure 8.4 The *Cat's Eye Nebula*. This is an example of a planetary nebula

After about 100 000 years the nebula has spread out so much that it fades from view and in the process mixes with the ISM, returning some of the star's original mass but with changes in its chemical composition.

The important point to understand is that stars are factories where elements are made. Whether low or high mass, successive nuclear reactions 'cook' nuclei inside a star's interior, creating many different elements in the process. The formation of the chemical elements in this way is called **stellar nucleosynthesis** and is the reason why they exist as they do in the periodic table today. Low mass stars cannot synthesise carbon into heavier elements such as silicon, iron, gold or uranium. The nuclear reactions in their core stop when carbon has been formed.

For high mass stars, the heavier elements up to iron can be made by nuclear fusion; elements beyond iron, but lighter than lead, can only be made by neutron capture.

Now that no more thermonuclear reactions are possible and all the star's fuel is used up, what happens next? What is the fate of a high mass star? The answer is that the star is now ready to die and the ways in which it can do this are discussed next in Chapter 9.

Summary

◆ Stars do not remain constant but change their luminosity and temperature with time. This information can be represented on a Hertzsprung–Russell Diagram (HRD). During its life, a star goes through four main evolutionary stages: **protostar, pre-main sequence star, main sequence star** and **post-main sequence star.**

◆ Whether or not a protostar makes it to the main sequence depends on the mass of the collapsing gas cloud during the process of star formation. If the mass of the cloud is less than $0.002M_\odot$ then the object is classified as a planet; if it is between $0.002M_\odot$ and $0.008M_\odot$ then it is called a **brown dwarf**. Brown dwarf stars shine weakly due to Kelvin–Helmholtz contraction and never develop core temperatures high enough for nuclear reactions to begin; consequently they never reach the main sequence.

◆ The Hertzsprung–Russell Diagram enables us to represent different stages in the life of a star. The path on the HRD followed by a star during its evolution is called its **evolutionary track**. At the point where a star first begins to shine by hydrogen burning and joins the main sequence, it is called a **zero-age main-sequence (ZAMS)** star. This marks the beginning of the longest phase of life during which all the energy it emits comes from converting hydrogen to helium in its core. The time a star spends on the main sequence depends on its mass. Hot massive stars spend less time than less massive cooler ones.

◆ When a star has used up all its hydrogen fuel, it will expand in size and become a **red giant**. Its luminosity increases and its surface temperature falls. It moves off the main sequence and over to the right on the HRD. **Helium burning** then begins in the star's core which involves a chain of nuclear reactions called the **triple alpha process** in which carbon is formed as an end-product.

◆ The production of heavy elements inside stars is called **stellar nucleosynthesis**. Nuclear reactions stop after carbon is formed in low mass stars, less than $3\,M_\odot$, and a **planetary nebula** is formed. High mass stars reach higher temperatures in their cores and can synthesise heavier elements, up to iron, by fusion reactions involving **carbon, neon, oxygen** and **silicon burning**. Beyond iron, fusion is no longer possible and the capture of neutrons by the **r- and s-processes** leads to the manufacture of elements heavier than iron.

Questions

1 What are the four main evolutionary stages in the life of a star?

2 A main sequence star with mass around 10 times the mass of the Sun will, at the end of its main sequence stage, start producing and radiating energy at a much higher rate. But its surface cools down. How can these two processes (faster energy radiation and cooling of the surface) happen at the same time?
What is the star called in this later stage?

ULEAC, Jan 1999, part

3 Sketch and label an HR diagram showing the main sequence and the regions occupied by *red giants* and *white dwarfs*.
Explain why main sequence stars of large mass have higher luminosities and shorter lives than main sequence stars of low mass.

UCEAC, Jan 1997, part

4 Calculate the following main sequence lifetimes in years of stars with the following masses and luminosities:

Mass/$M_\odot$	**Luminosity/$L_\odot$**
25	80 000
15	10 000
3	60
1.5	5
1.0	1
0.75	0.5
0.50	0.03

Explain how the Sun is expected to evolve. How does its time on the main sequence compare with stars of higher mass?

5 a What is meant by *stellar nucleosynthesis*?
 b What is the net energy released in the triple alpha reaction $3\,{}_{2}^{4}\text{He} \longrightarrow {}_{6}^{12}\text{C}$?
  [Note: ${}_{2}^{4}\text{He}$ has a mass of 4.0026 u and ${}_{6}^{12}\text{C}$ has a mass of 12.0000 u]
 c How are elements heavier than iron synthesised?

6 Compare and contrast the interiors of:
 a an old low mass star $< 3\,M_\odot$
 b an old high mass star $> 3\,M_\odot$

The deaths of stars

What happens to a star when it runs out of nuclear fuel? In this chapter we will see that the ultimate fate of a star depends on its mass. Low mass stars die in a continuous shedding of the star's outer layers, ejecting much of their mass into the ISM in the form of planetary nebulas. High mass stars destroy themselves as supernovas which are some of the most violent events in the universe. In both cases the ISM is enriched by a wide variety of heavy elements. What remains are three types of exotic stars: white dwarfs, neutron stars and the strangest of all – black holes.

White dwarfs

Over to the bottom left of the HRD (see Figure 5.5) is the region containing **white dwarf** stars. How are these stars produced and what is their composition?

Remember that in Chapter 8 we saw stars go through a series of nuclear fusion reactions, starting with hydrogen and then helium burning. In a low mass star, less than $3M_\odot$, as much as 25 to 60% of the star's mass is ejected in the form of a planetary nebula. After helium burning is complete, the temperature in the core is not high enough for the fusion of carbon and oxygen nuclei (helium burning's 'ashes') to take place, the star's temperature drops and its luminosity decreases. On the HRD, the evolutionary track of a low mass star crosses over the main sequence and then turns sharply downwards towards the white dwarf region (Figure 9.1).

Since nuclear burning has ceased, there is no longer any outward pressure to halt the crushing force of gravity. As a result, the core is compressed to a size roughly the same as the Earth and the density of matter rises to some 10^8–10^9 kg m^{-3}. To give you some idea of just how dense this is, a teaspoonful of white dwarf matter would have a mass of several tonnes!

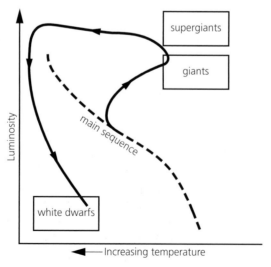

Figure 9.1 Evolutionary track of a star less than $3M_\odot$ into the white dwarf region on the HRD

Degeneracy

The star is prevented from further collapse by **degeneracy pressure**. To understand what this means we need to make use of a rule that arises from the quantum nature of matter. Particles of matter such as electrons, neutrons, protons and neutrinos obey a quantum rule called the **Pauli Exclusion Principle**. This, loosely stated, means that no two particles can exist together in exactly the same quantum state at the same time or, alternatively, that two particles cannot occupy the same space simultaneously.

A star in its normal main sequence phase can be modelled more or less as an ionised ideal gas obeying an **equation of state** which links the gas pressure P with the number of moles n of the randomly moving gas particles, and the temperature T of the gas. This equation of state is written as:

$$PV = nRT$$

where k is Boltzmann's constant $= 1.38 \times 10^{-23}$ J K^{-1} and R is the universal gas constant $= 8.3$ J mol^{-1} K^{-1}. However, when the gas is compressed to the very high densities reached in the interior of a white dwarf, the equation of state no longer holds. The particles cannot move randomly about and are tightly squeezed together to the extent that the electrons in neighbouring atoms overlap. The Exclusion Principle says that the electrons cannot be compressed any closer together, and as a result they exert a powerful outward pressure that opposes any further contraction by gravity. A gas in this state is called a **degenerate gas** and astrophysicists say that the core of the white dwarf is held together by **electron-degeneracy pressure**.

Unlike an ideal gas, the pressure exerted by a degenerate gas *does not depend on its temperature but on its density*, and heating it up or cooling it down makes no difference to the pressure exerted by it.

The Chandrasekhar limit

Stellar models show that the radius of a white dwarf is inversely proportional to the cube root of its mass, i.e.

$$R \text{ (white dwarf)} \propto \frac{1}{\sqrt[3]{M}}$$

This means that the more massive a white dwarf is, the smaller it becomes. This statement is true up to a certain limit called the **Chandrasekhar limit**. In 1931 Subrahmanyan Chandrasekhar (Figure 9.2), an Indian astrophysicist, showed that that the maximum amount of mass that a white dwarf can have and still be supported by electron-degeneracy pressure is about 1.4M$_\odot$. We could expect, therefore, all white dwarfs to have masses equal to or less than 1.4M$_\odot$. White dwarfs are much too faint to be seen directly with the naked eye. However, they can be observed by the effect they have on a companion star (see Box 9.1) and over 1000 have been detected to date.

Figure 9.2 Subrahmanyan Chandrasekhar (1910–95) who discovered the maximum possible mass for a white dwarf, now known as the Chandrasekhar limit

Box 9.1 Observing a white dwarf – *Sirius B*

A white dwarf can be detected by the effect it has on a companion star. A good example of this is the star *Sirius* in the constellation *Canis Major*. *Sirius* is the brightest star in the sky with apparent magnitude -1.4 and spectral type A0. it lies at a distance of 8.6 ly and is an early main sequence star.

As far back as 1844, using parallax measurements, it was noticed that the proper motion of *Sirius* was such that the star appeared to 'wobble' back and forth as if it were being influenced by some unseen object. It was soon deduced that *Sirius* was part of a binary system and telescopes revealed its white dwarf companion, *Sirius B*, shown at the 9 o'clock position (arrowed) in Figure 9.3.

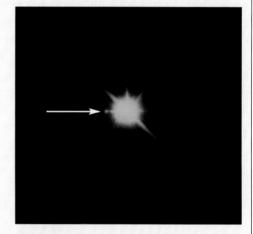

Figure 9.3 *Sirius B*, the white dwarf companion of *Sirius*

The colour of *Sirius B* enabled estimates of its surface temperature and radius to be made and it is found to have a mass of about $1.0M_\odot$ and a radius of $0.07R_\odot$. Its surface temperature is about 30 000 K so it is still quite hot and radiating.

Most white dwarfs are hard to see because they have cooled off and therefore have low luminosities. Over 1000 white dwarfs have been discovered, and it is thought that they may make up as much as 10% of all stars in our galaxy.

Composition of a white dwarf

What is the interior of a white dwarf like? As well as being very dense, as the star cools down the material forms into a crystalline lattice structure composed of ionised atoms with degenerate electrons moving freely within it. The star moves from a gaseous through to a liquid state and then becomes 'solid', and in many ways ends up with properties not unlike an electrically conducting metal such as copper or silver.

As thousands of millions of years pass, the white dwarf becomes colder and dimmer until its surface temperature approaches that of absolute zero, ending its life as a burnt-out cinder, sometimes called a **black dwarf**. This will be the fate of our Sun, ending its life as a dark ball of oxygen and carbon, but don't worry – the Sun is only about halfway along the main sequence so has plenty of life in it yet!

Neutron stars

For stars whose mass at the end of their thermonuclear burning phase is greater than the Chandrasekhar limit of $1.4M_\odot$ (but less than $3M_\odot$) not even electron-degeneracy pressure can prevent further collapse. The core temperature rises to some 10^9 K and the density to 10^{13} kg m^{-3}. Gravitational contraction is so strong that electrons are pushed into protons forming neutrons and releasing neutrinos by the process of **inverse beta decay**:

$$p^+ + e^- \longrightarrow n + \nu$$

and in about $\frac{1}{4}$ of a second the density of the core becomes the same as the density of an atomic nucleus, i.e. 2×10^{17} kg m^{-3}.

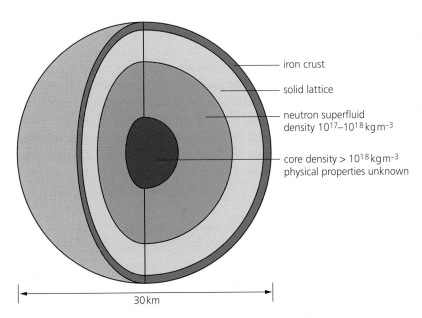

iron crust

solid lattice

neutron superfluid
density 10^{17}–10^{18} kg m^{-3}

core density $> 10^{18}$ kg m^{-3}
physical properties unknown

30 km

Figure 9.4 Simplified diagram of the interior of a neutron star

Figure 9.4 shows the interior structure of such a **neutron star**. At nuclear densities the particles form a **superfluid gas** composed of 80% neutrons, 10% electrons and the remainder protons (a superfluid is a fluid in which particles can flow over each other without friction). Surrounding the neutron star is a crust of iron covering a solid lattice composed of neutrons and neutron-rich nuclei. The superfluid neutron core is degenerate in the same sense as the electrons in a white dwarf are, and **neutron-degeneracy pressure** provides an outward force preventing any further collapse of the core, enabling the neutron star to become stable.

Neutron stars have strong magnetic fields because the magnetic field of the original star is now concentrated over a much smaller surface area, where the field strengths can now be as high as 10^8 T. Since the neutron star is so dense, it also has a very high surface gravity, some 10^{11} times that of the Earth. Its surface is very smooth – you won't find any high mountains on a neutron star!

To escape from a neutron star's surface you would need to travel at about 80% of the speed of light. Conversely, an object falling onto a neutron star would have at least this velocity and consequently an enormous amount of kinetic energy. This point is important when we come to discuss the origin of some of the highly energetic events recorded by astronomy satellites.

Supernovas

The formation of the neutron star happens extremely rapidly and once neutron-degeneracy pressure is established the core becomes rigid. The collapse of the star is dramatically halted and the infalling material bounces off the core and starts moving up towards the star's surface. A shock wave of tremendous energy is generated moving at high speeds (5000–10 000 km s^{-1}) and ejects the rest of the star's outer layers. The neutrinos produced by inverse beta decay swiftly travel out of the core, carrying away some 100 times more energy than is emitted as electromagnetic radiation. This gigantic explosion is called a **supernova** and can produce enough energy to temporarily outshine a whole galaxy!

How much energy is produced? The gravitational potential energy (GPE) of two bodies of masses m_1 and m_2 separated by a distance r is given by GPE $= -Gm_1m_2/r$. The negative sign indicates that it *increases* (or gets less negative) as the separation of the bodies gets larger, just as your GPE increases when you climb a mountain. If you jumped off the mountain it would all be converted to kinetic energy. Now consider the collapse of a star of mass $4M_\odot$ to a neutron star of radius $r = 15$ km. The loss of GPE will be:

$$\frac{(6.67 \times 10^{-11}) \times (8 \times 10^{30}) \times (8 \times 10^{30})}{1.5 \times 10^4} \text{ J} \approx 3 \times 10^{46} \text{ J}$$

which is much more than the energy of 10^{44} J normally observed in supernova events. So we can see that the formation of a neutron star easily provides enough energy to power a supernova explosion. Supernovas are truly among the most violent events in the universe!

Types of supernovas

Astronomers have found that it is possible to categorise supernovas into two types. The classification is based on the measurement of their light curves which are plots of their visual magnitude over time. **Type I** supernovas have a sharp maximum brightness of about $10^{10}\ L_\odot$ which then decreases gradually (Figure 9.5a). **Type II** supernovas have peak luminosities of about $10^9\ L_\odot$ and decline more rapidly (Figure 9.5b). The difference between the two types is thought to be due to two very different mechanisms producing the explosion.

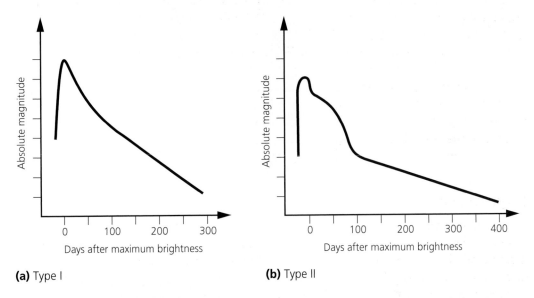

(a) Type I   **(b)** Type II

Figure 9.5 Supernova light curves

Type I are believed to be the result of a detonation in a close binary system involving a red giant and a white dwarf close to the Chandrasekhar limit. The high surface gravity of the white dwarf attracts gas from the red giant and pushes it over the limit of $1.4\ M_\odot$. The white dwarf violently collapses and its temperature increases, igniting rapid carbon burning and heavier element burning up to iron. The heat produced ejects matter from the star destroying it completely, leaving no neutron star behind.

Type II are due to the formation of neutron stars and the subsequent 'core bounce' of infalling matter that we have previously discussed. In Type I, the energy generated is due to thermonuclear processes, whereas in Type II the energy source is gravitational. In both cases heavier elements are synthesised and ejected into the interstellar medium.

Astrophysicists are still unsure of the exact processes that give rise to supernovas; and what has been described is based on computer models of stellar structure. Supernovas are not frequently seen, but an unprecedented opportunity to observe one at relatively close range presented itself in 1987 (see Box 9.2).

Box 9.2 Supernova 1987A

Figure 9.6 Supernova 1987A. The image on the left shows the position of the progenitor star *Sanduleak 202–69* (arrowed). The right shows the sudden increase in brightness of the supernova

On 24th February 1987 a supernova event occurred in the *Large Magellanic Cloud* (LMC) which was visible to the naked eye (Figure 9.6). The LMC is a small neighbouring galaxy about 170 000 ly away which means that the supernova (designated SN 1987A as it was the first supernova observed in the year 1987) exploded 170 000 years ago.

SN 1987A is the first supernova in recent times to be observed relatively close to the Earth using modern instrumentation. Also, the nature of the star preceding it (or the **progenitor**) called *Sanduleak 202–69* was known. One mystery is that *Sanduleak* was a *blue* and not a red supergiant star. The solution to this puzzle was that *Sanduleak* was initially a massive star of at least $20M_\odot$ which had lost mass and whose evolutionary track had brought it back to the blue side of the HRD before its iron core collapsed. About one day before the supernova became visible, neutrino bursts were detected in the Kamiokande II neutrino detector in Japan and also from detectors in a salt mine in Ohio, USA.

The light curve was somewhat different from that expected for Type II supernovas. SN 1987A did not get as bright as anticipated and its peak brightness remained more or less constant for 40 days before declining gradually and then more steeply over a period of about two years.

The Hubble Space Telescope has photographed three rings of gas surrounding SN 1987A (Figure 9.7). These rings are believed to be material ejected from *Sanduleak* by stellar winds shortly before the explosion occurred.

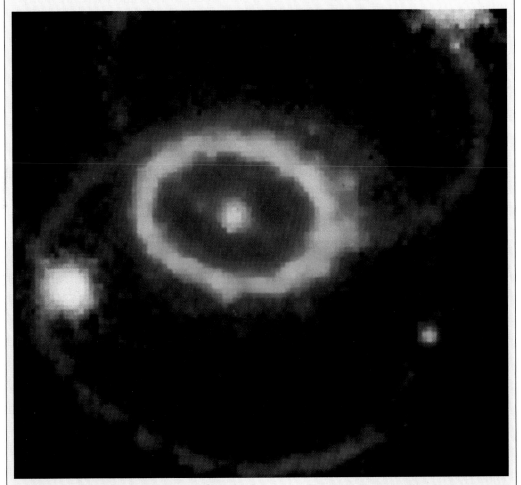

Figure 9.7 Rings of material ejected from *Sanduleak* before the supernova event

The arrival of the neutrino bursts confirms at least some aspects of supernova models, and for the first time we may have witnessed the formation of a possible neutron star through inverse beta decay. Calculations show that SN 1987A released about 10^{46} J of energy, which is also consistent with theory.

Supernova remnants

In AD1054 Chinese astronomers recorded the appearance of a 'guest star' that remained visible during the *day* for some three weeks! At night it was visible to the naked eye for a further 650 days. We now know that what the Chinese astronomers saw was a supernova event. Figure 9.8 shows the **supernova remnant** that we see today. It is in the form of a large expanding shell of gas called the *Crab Nebula* and lies in the constellation of *Taurus*.

Supernova remnants are the debris of supernova explosions and there are many scattered across the sky. Photographs show them as clouds of glowing gas several parsecs across. For the nearer remnants, it is possible to measure the expansion rate of the shell and therefore deduce the date of the original explosion. This is how we can be confident of the association of the *Crab Nebula* with the Chinese astronomers' record. The expanding gas cloud slams into the interstellar medium at high velocity causing atoms in the gas to become excited and emit electromagnetic radiation over a wide range of wavelengths from radio through to X-ray.

Supernovas have been observed in galaxies other than our own, and based on these observations, astronomers estimate we should expect to see, on average, about five supernova events per century. The fact that we see fewer than this in our own galaxy is probably due to obscuring clouds of interstellar gas and dust lying between us and the supernovas.

Figure 9.8 The *Crab Nebula*. This is an example of a supernova remnant

Novas – 'small' stellar explosions

Stellar explosions that occur more frequently but are less energetic are called **novas**. A nova is observed as a star that suddenly increases in luminosity by as much as $10^5\, L_\odot$ and stays at this level for a few hours before decreasing sharply and gradually over a period of several hundred days. The energy released is not nearly so great as that from a supernova – 'only' some 10^{37} to 10^{38} joules! So what mechanisms power a nova event?

Observations of novas show that they all seem to occur in binary star systems with short orbital periods. The stars are so close together that material can flow from one to the other through a region called the **Roche Lobe**. At the Roche Lobe the gravitational attraction of material from one star exactly balances that from the other, enabling matter to flow between them (Figure 9.9).

One model for a nova involves a red giant and a white dwarf in orbit around each other. As the red giant expands in size, matter reaches the Roche Lobe and then falls onto the white dwarf. As the matter spirals in under the white dwarf's gravity, it forms a disk of material called an **accretion disk**. A disk is formed as a consequence of the **conservation of angular momentum** (see Box 9.3, page 170) – the rotation of the star causes the material to be spread into a disk, rather like when you pour cream into coffee (having first stirred it).

The material that moves onto the white dwarf's surface is mainly hydrogen which forms an envelope around the star. As more hydrogen is collected it compresses the underlying layers, heating them to temperatures exceeding 10^6 K when explosive hydrogen burning ignites the envelope; this is what we observe as a nova. Another model of a nova event involves a main sequence star that sheds mass onto a white dwarf by stellar winds. This moves onto the white dwarf in the same way, triggering a nova explosion.

Stellar models show that only a relatively small mass infall is needed to cause a nova event. Some studies have shown that as little as $10^{-13} M_\odot$ of material accreting onto the white dwarf is sufficient and some novas may even be single white dwarfs accreting matter from the ISM.

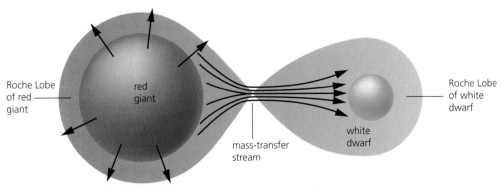

Roche Lobe of red giant — red giant — white dwarf — Roche Lobe of white dwarf — mass-transfer stream

red giant expands to the Roche Lobe and matter can flow onto the white dwarf

Figure 9.9 The Roche Lobe – this is where matter can flow easily from one star to another

X-ray bursters

During the early years of X-ray astronomy, it was discovered that some X-ray objects emitted brief but very energetic bursts of X-rays lasting for several seconds before dying down. These sources are known as **X-ray bursters** and flare up in intervals ranging from several days to a few seconds with some even shooting off bursts of energy like a machine gun!

Measurements using X-ray astronomy satellites such as ROSAT have found bursters to have peak X-ray luminosities of the order of 10^{31} W. By analysing the X-ray spectrum astronomers can deduce that the temperature required to produce these bursts is about 3×10^7 K but what sort of object could be responsible?

A clue is found using the Stefan–Boltzmann relation $L = A\sigma T^4$ (see Chapter 2). Remembering that stars are good approximations to blackbodies, putting in the values for luminosity and temperature we find that the surface area A of the object is:

$$A = \frac{L}{\sigma T^4} = \frac{10^{31} \text{ W}}{(5.67 \times 10^{-8} \text{ W m}^{-2} \text{ K}^{-4}) \times (3 \times 10^7 \text{ K})^4} \approx 2 \times 10^8 \text{ m}^2$$

and assuming that the object is spherical, $A = 4\pi r^2$, so its radius r is found to be:

$$r = \sqrt{\frac{A}{4\pi}} = \sqrt{\frac{2 \times 10^8 \text{ m}^2}{4\pi}} \approx 4 \times 10^3 \text{ m}$$

i.e. the object has a diameter of 8 km which is about the same size as a neutron star. The accepted model of an X-ray burster does indeed involve a neutron star, and the mechanism is similar to that of a nova except that we substitute a neutron star for a white dwarf. Hydrogen from a companion falls onto a neutron star by accretion. As it does this, the hydrogen becomes so hot that hydrogen burning by the CNO cycle begins before it reaches the star's surface and a helium envelope builds up where the temperatures increase to some 2×10^9 K. Rapid helium burning ignites the neutron star in a helium 'flash' that releases vast amounts of X-rays in a few seconds. As more helium is accreted, the process repeats itself until all the matter from the companion star is used up.

Neutron stars in binary orbits are very dynamic systems and can explain a number of other X-ray phenomena. Recall what we said about the high velocity that matter has when falling onto a neutron star. A pulsating X-ray source can be explained by infalling matter being funnelled down to the neutron star's magnetic poles where it strikes the surface with such high kinetic energy that it forms 'hot spots' which release beams of X-rays in opposite directions. If the neutron star is rotating and the Earth is in the line of sight, then it will look like a regular pulsating source. The 'machine gun' effect is thought to be many smaller hot spots going off at once. Even as the gas accretes it will heat up to give a certain amount of continuous X-ray emission contributing to a background from which the intensity increases.

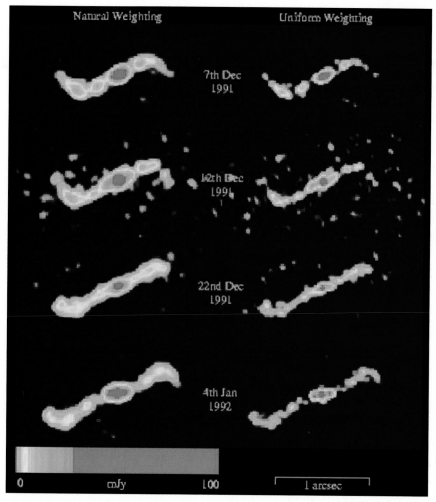

Figure 9.10 (a) SS 433. This is a radio image that has been image processed to bring out detail. Notice the scale calibrated in mJy, which indicates the strength of the radio emission across the object

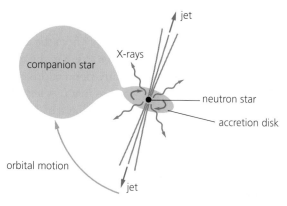

Figure 9.10 (b) A possible model for SS 433

Neutron stars have also been observed ejecting matter. Figure 9.10a shows a series of radio images, taken by the MERLIN Very Large Baseline radio interferometer in the UK, showing jets of matter being ejected at high velocity from a star called SS 433. Figure 9.10b shows a possible model. SS 433 is a neutron star in close binary orbit with a companion and is collecting matter from it. However, the collection rate is so high that pressure builds up along the plane of the accretion disk and is eventually relieved by the expulsion of large amounts of matter streaming along the path of least resistance, which is parallel to the rotation axis of the disk.

Box 9.3 Angular momentum

The **linear momentum** of a mass m travelling with a uniform velocity v is given by $m \times v$ and is measured in kg m s^{-1}. The total linear momentum of an isolated system of particles is *conserved*, i.e. the total momentum of the system remains constant.

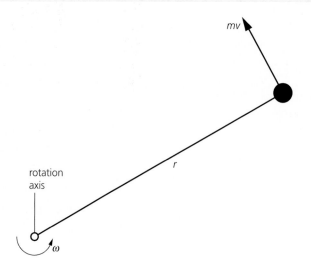

Figure 9.11 Angular momentum $L = mvr = mr^2\omega$

If a point mass m is rotating about an axis at a distance r from the axis, we can define a vector quantity called the **angular momentum** as $L = m \times v \times r$, the momentum where $m \times v$ is the tangential momentum at any point. An alternative form of this equation is found by putting v in terms of the angular velocity ω as $v = r\omega$, so that $L = mr^2\omega$ (Figure 9.11). For a rotating extended body, the angular momentum can be thought of as the sum of the elemental angular momentum of each part of it about the rotation axis. Angular momentum has units of kg m^2 s^{-1}.

Like linear momentum, angular momentum is also conserved and the **law of conservation of angular momentum** states that the total angular momentum of an isolated system of particles rotating around a common axis remains the same.

You may have seen the conservation of angular momentum in action by watching a spinning ice skater. As the skater pulls her arms towards her body, she starts spinning much faster. This happens because her angular momentum L is constant, so as the effective radius of her arms becomes less as she folds them in, her tangential velocity has to increase to keep L the same.

Angular momentum and moments of inertia

The conservation of angular momentum requires that neither the magnitude nor the direction of L changes throughout the motion. This means the value of L is constant and the axis of rotation remains fixed in space. Since rotating objects have different shapes (e.g. a disk or a sphere), it is useful to define another quantity called the **moment of inertia** I which expresses the way in which mass is distributed throughout a rotating object. We can then write the angular momentum in terms of the moment of inertia as $L = I\omega$. The moment of inertia for a sphere of mass M and radius R is given by $I = \frac{2}{5}MR^2$.

Angular momentum is important in astrophysics since many astronomical objects rotate. The Earth rotates as do the Sun and other stars. The galaxy is also rotating, and we can use the conservation of angular momentum to find out information about the physical properties of astronomical objects. One example is a rapidly spinning neutron star, which is the result of the sudden collapse of a star in a supernova event.

Suppose a star of radius equal to that of the Sun (6.96×10^5 km) suddenly collapses to a neutron star of radius 12 km. What would be the angular speed of the neutron star?

Assuming that there is no mass lost in the process, let the initial angular speed of the star be ω_{star} and that of the neutron star ω_{neut}. If angular momentum is conserved:

$$L_{star} = L_{neut}$$

$$\tfrac{2}{5}MR^2_{star}\omega_{star} = \tfrac{2}{5}MR^2_{neut}\omega_{neut}$$

so
$$\left(\frac{\omega_{neut}}{\omega_{star}}\right) = \left(\frac{R_{star}}{R_{neut}}\right)^2$$

Putting in the respective values of the radii:

$$\frac{\omega_{neut}}{\omega_{star}} = \left(\frac{6.96 \times 10^5 \text{ km}}{12 \text{ km}}\right)^2 = 3 \times 10^9$$

In other words the neutron star would spin some 10^9 times faster than the star from which it was formed! In reality, the neutron star would not be rotating as fast as this, since a significant amount of matter from the star would be ejected into space from the supernova explosion and some of this would carry away angular momentum.

Pulsars

In 1967 at the Mullard Radio Astronomy Observatory at Cambridge, a graduate student Jocelyn Bell, now Professor of Physics at the Open University, was working with an experimental radio telescope when she noticed that the receiver was detecting regular radio pulses from a specific area in the sky. The radio pulses were extremely regular with a period of 1.337 301 13 seconds, and many more were soon discovered with periods ranging from 0.25 to 1.5 seconds. These objects were called **pulsars** (for *pulsating* radio sources). But what was causing these regular radio transmissions? The mystery was solved when a pulsar was discovered in the centre of the *Crab Nebula*. We know that the *Crab Nebula* is the remains of a supernova event that occurred in 1054, and that neutron stars are formed when massive stars explode. Pulsars are rapidly spinning neutron stars and Figure 9.12 shows a model which can explain their behaviour.

In our discussion of neutron stars we mentioned that they have very high magnetic fields. On the surface of the neutron star are numerous protons and electrons, as the gravitational field strength at the surface is not strong enough for

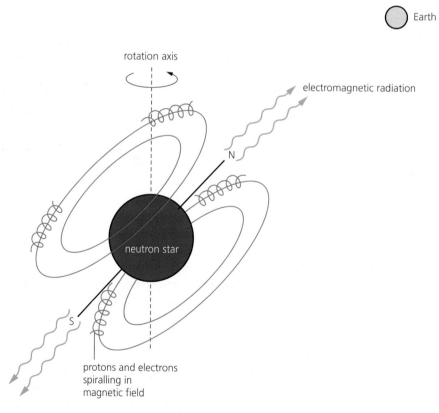

Figure 9.12 Lighthouse model of a pulsar. Protons and electrons spiralling in the pulsar's strong magnetic field produce two opposing narrow beams of electromagnetic radiation

inverse beta decay to occur. These are accelerated towards the magnetic poles and in doing so emit electromagnetic radiation over a wide range of wavelengths in a narrow beam in opposite directions.

A neutron star rotates rapidly because of the law of conservation of angular momentum (see Box 9.3). All stars rotate to a greater or lesser extent, but when reduced in size they start spinning faster. As the star rotates the beams of radiation swing round with it rather like a beam of light from a lighthouse. It just so happens that the pulsar in the *Crab* is aligned at such an angle so that we get a good view. The *Crab Nebula* pulsar is one of the fastest ever discovered with a period of 33 milliseconds and, as well as the emission at radio wavelengths, visible radiation has also been detected.

Another interesting feature of some pulsars is that they can tell us something about the structure of the neutron star. Accurate timing measurements show that sometimes the period speeds up in what astrophysicists call a 'glitch'. For this to happen the solid crust of the star must be brittle, and as it cools and settles, the diameter of the neutron star alters slightly and because of the conservation of angular momentum its rotation speed 'flinches'. In other words the crust experiences 'starquakes' analogous to the earthquakes we experience due to movement of the Earth's crust.

White dwarfs and neutron stars are examples of what astrophysicists call **compact objects**. These are objects that are small, have high densities and low luminosities. We saw that the Chandrasekhar limit determines the maximum mass of a white dwarf star and there is also an upper limit to the mass of a neutron star, which stellar models indicate lies between 2.5 and $3M_\odot$. But what would happen if the mass of a stellar corpse exceeded the limit for neutron stars? Not even neutron-degeneracy pressure would be able to resist the gravitational compression. To answer this question, astrophysicists have predicted the existence of one of the most exotic and bizarre objects in modern astrophysics – a **black hole**.

Black holes

To understand what a black hole is we first have to look at the some of the work of Albert Einstein, particularly his **General Theory of Relativity** and how it explains on a universal scale what gravity is. We will discuss the general theory in more detail in Chapter 10, but here are three key results from general relativity that are relevant to the astrophysics of black holes:

Result 1 Unlike Newtonian mechanics, space and time are not separate entities but are linked together. In general relativity, in order to describe an event completely, we specify three co-ordinates in space and add a fourth one of time. These four co-ordinates describe the points of **space–time** in which particles of matter move.

Result 2 Space–time is *curved* in the presence of matter. The distribution of mass determines the degree of curvature. As a result, particles of matter move along curved paths. The geometrical path they take is what we call the 'effect of gravity'. Unlike Newtonian mechanics, the concept of gravity as 'force' (i.e. action at a distance) is not used in general relativity; here gravity is the distortion of four-dimensional space–time.

Results 3 One important consequence of the curvature of space–time is that the presence of mass causes light to deviate from a straight line. The more mass there is, the more the light path is curved. This phenomenon is fundamental to the definition of a black hole.

The effects of gravity as far as general relativity is concerned may be summed up by the ideas that '*space–time tells matter how to move*' and '*matter tells space–time how to curve*' (see Figure 10.3, page 191).

The formation of black holes

How is a black hole formed? Early in this chapter we explained how a collapsing star can be supported by degeneracy pressure; electron-degeneracy in the case of a white dwarf and neutron-degeneracy for a neutron star. For very massive stars, however, no such mechanism exists to halt the relentless compression by gravity. For a star with a main sequence mass greater than $10M_\odot$ gravity squashes it to such an extent that, in theory, its density becomes infinite and its volume zero! This state of matter is called a **singularity** and is inaccessible to the laws of physics as we understand them.

The gravitational field surrounding a black hole is so high that *no* radiation (including light) can escape and as a result it appears black due to the absence of any directly observable emission.

The Schwarzschild radius

How big is a black hole? To answer this question, consider the amount of energy needed by a mass m to escape to 'infinity' from the surface of a body mass M and radius R. From astrodynamics, the escape velocity is given by $v_{esc} = \sqrt{2GM/R}$ (see Chapter 2, page 44). The maximum escape velocity that it is possible to have is c since nothing can travel faster than the speed of light, and so

$$R = R_{Sch} = \frac{2GM}{c^2}$$

This radius is called the **Schwarzschild radius** after the German astrophysicist Karl Schwarzschild (1873–1916), who first calculated it from the General Theory of Relativity. The Schwarzschild radius tells us, for a given mass, how small an object must be for it to trap light and therefore appear black.

The event horizon

Figure 9.13 shows the structure of a black hole. The region of space where the escape velocity from the black hole is equal to the speed of light is called the **event horizon** and appears from outside the black hole to have a radius equal to R_{Sch}. Inside the event horizon the escape velocity required would exceed that of light and so once inside it there is no escape!

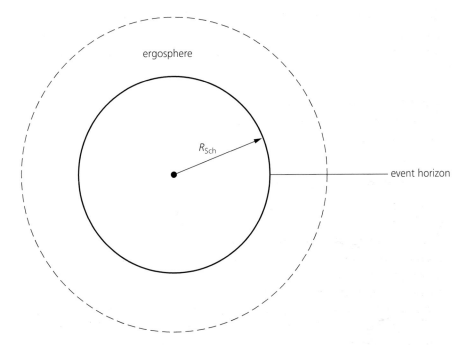

Figure 9.13 Structure of a black hole. The ergosphere is a region of distorted space–time outside the event horizon which is dragged around the black hole as it rotates

The physical properties of black holes

When an object has passed the event horizon, we cannot know anything further about its existence so a black hole effectively removes information such as matter and energy from the universe. For a star that has collapsed past the event horizon into a black hole, information about its chemical composition, colour, shape and size are lost forever. It is, however, possible to describe all black holes by just three physical quantities: their *mass*, *electric charge* and *angular momentum*.

Mass

In theory, if we were in orbit in a spaceship around a black hole (keeping away from the event horizon of course), we could use Kepler's 3rd Law to determine its mass by measuring the size and period of our orbit.

Electric charge

Since the electric force, like gravity, is a long range interaction, we could measure from our spaceship the electric field strength that the black hole has and from this determine its charge using Coulomb's Law.

Angular momentum

Due to the conservation of angular momentum we expect the black hole to be rotating. General relativity predicts that this rotation will cause space–time to be dragged around with the hole in a region called the **ergosphere** which lies outside the event horizon (see Figure 9.13). By observing the motion of objects moving in the ergosphere it would be possible to determine the black hole's angular momentum.

Inside a black hole

So what would conditions be like inside a black hole? We have already said that once the event horizon has been crossed there is no way of getting back. A hapless astronaut would be drawn inexorably towards the singularity. A prediction of general relativity is that all gravitational fields cause time to slow down. From our spaceship situated far from the horizon, we would observe our astronaut colleague fall into the hole at a slower and slower rate until, at the event horizon, he would seem frozen in time. From his own point of view he passes the event horizon accelerating all the time. Gravity would exert differential forces on his length, tearing him apart as he crashes towards the singularity.

Inside the event horizon general relativity predicts that space and time become confused and the laws of physics actually break down at the singularity. The laws of physics, depending as they do on a well defined framework of space and the direction of time, become meaningless. However, since this confusion of physical reality is unable to communicate itself across the event horizon, the laws of physics in our observable universe remain unaffected. This bizarre state of affairs has been summed up by the British mathematician Sir Roger Penrose (1931–) as the law of cosmic censorship – '*Thou shalt not have naked singularities*'. In other words, singularities do not exist without an event horizon, in order to prevent them from interfering with the physical laws in our own universe.

Observing black holes

As a black hole does not shine like normal stars we can only infer the existence of one by the effect that it has on nearby objects. There are two ways in which we can do this.

X-ray sources

If a black hole forms a part of a binary system then matter falling into it will become very hot as it gains kinetic energy, forming an accretion disk and emitting electromagnetic radiation. As the temperature of the infalling gas reaches values in excess of 10^6 K it will emit X-rays that can be detected by X-ray astronomy satellites. If the distance to an object is known, then its X-ray luminosity can be determined using the inverse square law. In this way, X-ray luminosities have been detected for many objects ranging from 10^{26} to 10^{31} W (see Box 9.4). Using Wien's Law and Stefan's Law, we can then determine how big the object must be.

WORKED EXAMPLE 9.1

The ROSAT X-ray astronomy satellite detects an X-ray source of luminosity 10^{30} W at a peak wavelength of 0.3 nm. How big is it?

We suppose the object is a spherical blackbody with a radius R. For it to emit at electromagnetic radiation at X-ray wavelengths then using Wien's Law it must have a temperature of:

$$T = \frac{2.90 \times 10^{-3} \text{ m K}}{\lambda_{max}} = \frac{2.90 \times 10^{-3} \text{ m K}}{0.3 \times 10^{-9} \text{ m}} \approx 10^7 \text{ K}$$

From Stefan's Law its radius must therefore be:

$$R = \left(\frac{L}{4\pi\sigma T^4}\right)^{\frac{1}{2}} = \left[\frac{10^{30} \text{ W}}{4\pi(5.7 \times 10^{-8} \text{ W m}^{-2} \text{ K}^{-4})(10^7 \text{ K})^4}\right]^{\frac{1}{2}} \approx 12 \text{ km}$$

This is the size of a neutron star *or* it could be the radius of the accretion disk around a black hole.

Gravitational lensing

A black hole which is not in a binary system may sweep up matter from the interstellar medium which falls into it causing it to radiate. In this case it is not possible to identify the black hole with a visible companion although X-rays would be detected. Another way a black hole might be found is through the distortion it creates in space–time. Suppose that the Earth, a star and a black hole are almost aligned as shown in Figure 9.14. As light from the star passes close to the black hole it will be bent round either side of it due to the strong gravitational field. From Earth we should therefore see *two* images of the star, a brighter primary image on the side closest to the black hole and a fainter secondary image on the other side. (If the star, the black hole and the Earth were in perfect alignment then the intensity of the images would be the same.) This phenomenon has actually been observed, providing confirmation of the General Theory of Relativity. It is called **gravitational lensing** since the black hole acts like a lens, distorting the image of the star.

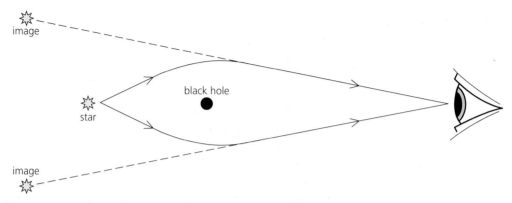

Figure 9.14 Gravitational lensing. The black hole deflects light from a star behind it so that the observer sees two images

Box 9.4 *Cygnus X-1* – a black hole?

X-ray astronomy satellites have found a strong source of X-rays in the constellation of *Cygnus*, the Swan. This object, called *Cygnus X-1*, has a luminosity at X-ray wavelengths of some 4×10^{40} W which varies over short time intervals of 0.001 s or less. When telescopes are pointed in the direction of *Cygnus X-1*, they reveal a blue supergiant star designated in star catalogues as HDE 226868, which lies at a distance of 8000 ly from Earth (Figure 9.15).

Supergiants do not emit large amounts of X-rays, so the X-ray emission must be coming from a smaller companion star of HDE 226868 which is invisible to us. Spectroscopic analysis of HDE 226868 shows that its spectral lines are Doppler shifted and move back and fourth with a period of 5.6 days, indicating that it is part

Figure 9.15 This blue supergiant HDE 226868 (arrowed) has an unseen companion that is probably a black hole

of a binary system. Using the mass–luminosity relation, the mass of HDE 226868 is estimated to be in excess of $20M_\odot$ which implies that the unknown companion must have a mass greater than $6M_\odot$ for it to exert any appreciable effect on the orbit of the supergiant star. This rules out a white dwarf or a neutron star whose masses would be too small, so the only other possible candidate for *Cygnus X-1* is a black hole. The X-ray emission is thought to come from material blown out from HDE 226868 by its stellar wind. This material is heated to high temperatures as it falls into the singularity, forming an accretion disk around the hole. The variation of X-ray intensity is probably due to the rapidly rotating accretion disk generating small concentrations of heated gas or 'hot spots' on its inner surface.

Hawking radiation

Finally, another property of black holes has been put forward by the British theoretical physicist Stephen Hawking (1942–), who predicts that they could actually evaporate! Hawking combined quantum theory with general relativity to show that near a black hole, some types of matter can appear out of the vacuum to create particle–antiparticle pairs. If one of these particles then falls into the black hole while its counterpart escapes, then we consider the gravitational potential energy of the black hole to have produced this pair and this is equivalent to the escaping particle carrying away some of its mass. An observer would then see a steady emission of particles coming from the black hole which is known as **Hawking radiation**. Hawking showed that for a black hole of mass M, the rate of energy loss is proportional to $1/M^2$ so that as the black hole's mass decreases, the 'evaporation' rate increases. As the mass dwindles to zero, the evaporation rate increases extremely rapidly, leading to a final explosive burst of elementary particles together with the emission of gamma rays.

Hawking calculated that the time for a black hole to evaporate t_{evap} is approximately:

$$t_{evap} \approx \left(\frac{M}{M_\odot} \right)^3 \times 10^{66} \text{ years}$$

This is a very long time and, for the range of masses of collapsing stars, is of no consequence since the age of the universe is estimated to be between 10 and 20×10^9 years (see Chapter 10). However, astrophysicists like Hawking believe that very early on in the history of the universe very small black holes (less than the size of an atom) with a mass of some 5×10^{11} kg could have formed. In this case t_{evap} would be about 15×10^9 years and so these primordial black holes ought to be completely evaporated about now. The high energy gamma rays emitted should be detectable by gamma ray astronomy satellites, and gamma ray bursts have indeed been discovered. Are some of these the final moments of black holes left over from the Big Bang?

Stardeath – a summary

To sum up, the various ways in which stars can meet their end and the relationship to their main sequence mass is shown in Table 9.1.

Table 9.1 The deaths of stars

Main sequence mass/*M*	Ultimate fate
0.1–0.5	White dwarf
0.5–4	Planetary nebula, then white dwarf
4–10	Supernova, then neutron star
10–20	Supernova, then neutron star *or* black hole
20–60	Supernova, then black hole

Summary

◆ The fate of a star after it has used up its nuclear fuel depends on its **mass** at its time of death.

◆ For stars with main sequence masses between 0.1 and $4M_{\odot}$, the end product is a **white dwarf** star and, for mass $0.5–4M_{\odot}$, a **planetary nebula** is also formed.

◆ A white dwarf star is prevented from further collapse by **electron-degeneracy pressure**. The maximum mass a white dwarf can have is determined by the **Chandrasekhar limit** of $1.4M_{\odot}$.

◆ For massive stars from 4 to $20M_{\odot}$, the Chandrasekhar limit is exceeded, causing the star to collapse into a **neutron star**. The formation of a neutron star results in the star violently exploding as a **supernova**. Neutron stars are prevented from further collapse by **neutron-degeneracy pressure**.

◆ Supernovas are classified into two types, **Type I** and **Type II**, based on the shape of their light curves. Type I are caused by the explosion of solar mass stars in a binary system, where one of the members is a white dwarf close to the Chandrasekhar limit. Type II are due to massive stars greater than $3M_{\odot}$ that through continuous stages of nuclear burning have evolved iron cores.

◆ A **nova** is a small-scale stellar explosion caused by matter from a companion star in a close binary system falling onto the surface of a white dwarf by a process of **accretion**. Novas are less energetic but occur more frequently than supernovas.

◆ **X-ray bursters** are the result of matter from a companion star in a close binary system accreting onto the surface of a neutron star. Matter can also be ejected perpendicularly from the plane of an accretion disk in some binary systems involving neutron stars. In both cases large amounts of X-rays are emitted due to heating of hydrogen and helium gas.

◆ **Pulsars** are rotating neutron stars which emit regular pulses of electromagnetic radiation. Pulsars were first detected at radio wavelengths but have also been observed optically. The radio emissions are generated by high speed electrons moving in the neutron star's high magnetic field.

◆ A **black hole** is an object that results from the death of a star with mass greater than $10M_{\odot}$. Black holes are so called because nothing, not even light, can escape from them. They can only be described by three quantities: mass, charge and angular momentum.

◆ Space and time are distorted in the vicinity of a black hole as described by Einstein's **General Theory of Relativity**. Black holes can be detected by the effects they have on nearby objects, such as the X-ray emission from gas falling into them, or by **gravitational lensing**.

Questions

1 What is meant by degeneracy pressure? What is the difference between electron-degeneracy pressure and neutron-degeneracy pressure? To what kinds of stellar objects do they apply?

2 A white dwarf has a mass of 1.4×10^{30} kg and a radius of 1.5×10^6 m.
  a Calculate the gravitational field strength at its surface.
  b Calculate the gravitational potential at its surface.
  c Using your answer to **b**, and that KE $= \frac{1}{2}mv^2$, calculate the escape speed for a body to escape from the surface of the white dwarf.

3 Calculate the kinetic energy of matter containing $2M_\odot$ ejected from a supernova event with a velocity of 5×10^3 km s^{-1}.

4 A star of radius 15×10^5 km rotates about once a month. It then explodes in a supernova leaving behind a neutron star of radius 10 km. Assuming no mass loss and using the law of conservation of angular momentum, find the angular velocity ω of the spinning neutron star. (1 month $= 2.6 \times 10^6$ s)

5 Using information from X-ray astronomy satellites, astronomers have found a very powerful X-ray source associated with a neutron star called GROJ 1744-28. Astrophysicists have deduced that it has a gravitational field approximately 10^{11} times that of the Earth and a radius of about 8 km. Estimate the mass of GROJ 1744-28. Assuming that GROJ 1744-28 is part of a binary system and is accreting matter, outline a mechanism for the X-ray emission.

6 Calculate the Schwarzschild radius for
  a the Earth
  b the Sun
  c the galaxy.

  What do you notice about your results? (Use the data given on page 220.)

7 a Give two ways in which black holes may be detected.
  b What is meant by the term *singularity*? In what way does a black hole impose 'censorship' on our universe?

8 This question is about gravity and some of the effects of gravity which are important in modern astronomy and which lead to the idea of a *Black Hole*.
    When a projectile is fired vertically from the surface of the Earth with a small velocity v, it reaches a height which depends upon v, and then falls back to the surface. Above a certain velocity, called the *escape velocity*, it does not return to the Earth but carries on indefinitely. The escape velocity is found by equating the projectile's initial kinetic energy to the gravitational potential energy gain in moving the projectile from its starting point to 'infinity'.

181

a Show that the escape velocity of a projectile from the surface of a spherical body of mass M and radius R is given by:

$$v_{esc} = \sqrt{2GM/R}$$

where G is the universal gravitational constant.

b Now explain **why** the escape velocity of the projectile can be found by equating its initial kinetic energy to its gravitational potential energy gain in moving from this starting point to 'infinity'.

c Using data for the Sun given on page 220, show that:
  i) the mass of the Sun is 2.0×10^{30} kg, and
  ii) the escape velocity from its surface is 6.2×10^5 m s^{-1}.

d In the later stages of a star's life, when the star has used most of its energy resources, it can shrink to a very much smaller diameter. For example, if the Sun were to shrink until its density were comparable to the density of an atomic nucleus it would have a radius of about 12 km. A star in this state is called a *neutron star*.
  Show that the escape velocity from the surface of the Sun, if it **were** to shrink to this size, would be close to one half of the speed of light.

e A Black Hole is an object with an escape velocity equal to the speed of light so that nothing, not even light, can ever escape from the surface of the object.
  What would the radius of the Sun have to be in order for it to become a Black Hole?

f Although theory tells us that Black Holes ought to exist, it is very difficult to prove that they do since they cannot be observed directly.
  Similarly, neutron stars are difficult to see since they are so small and very faint. However, many stars in the sky have close companions and this can be particularly helpful in making observations on neutron stars.
  If a neutron star has a companion which is very close and very large, the gravitational field of the smaller star can draw matter from the outer layers of the larger one. The gravitational field at the surface of the neutron star is so intense that the falling matter attains an extremely high velocity, becomes very hot when it reaches the surface, and so radiates sufficient electromagnetic energy to be detected by astronomers.
  Calculate the kinetic energy attained by one mole (2.0×10^{-3} kg) of hydrogen gas falling onto the surface of the Sun, if the Sun had a radius of 12 km.

g Now use this figure to estimate the temperature reached by the gas if all of that energy were used to heat the gas.

h A **very rough** idea of the energy of the most energetic photons emitted by a hot gas at a temperature T can be found by equating kT, where k is the Boltzmann constant, to the energy of a photon.
  i) Use this argument to estimate the **smallest** wavelength of a photon emitted by a gas at a temperature of 10^{12} K.
  ii) In what part of the electromagnetic spectrum does this radiation lie?

(Use the data from page 220; take the density of the Sun to be 1.4×10^3 kg m^{-3}.)

O & C, Physics (Nuffield), June 1988

Cosmology

'How did the universe begin?' and 'What will happen to it in the future?' are questions that have occupied the human mind since antiquity. However, it is only now late in the 20th century that we seem close to constructing a plausible explanation of how the universe was created and what its eventual destiny might be. By combining astrophysics with particle physics and using key observational data, cosmologists have developed a model of the universe which attempts to explain its origin and ultimate fate.

What is cosmology?

The study of the structure and development of the universe as a whole is called **cosmology** and scientists who study such things are called **cosmologists**. The task of the cosmologist today is to understand how different phenomena of nature, from small elementary particles and fundamental forces right up to very large-scale structures in the universe such as clusters of galaxies, all fit together.

Any theory of origins must take into account the observational evidence. We will start by looking briefly at some early ideas about the universe and how they accounted for the limited observational facts that were known at the time. Then we will see what evidence we have now, and make some key assumptions about the universe in general. Finally we will see how a theory of origins emerges that can be tested against the observational evidence and look at some still unanswered questions.

Early models of the universe

The Sumerians and the Babylonians

The earliest civilisation is believed to be that of the **Sumerians** (c.7000BC) who lived in the region of the Middle East known as Mesopotamia, in what is now called Iraq. The Sumerians are thought to have invented the art of writing using clay tablets with inscriptions in **cuneiform** characters and it is from these that we know they were a highly advanced culture. The Sumerians recorded the positions of the stars and planets and gave names to some of the constellations we know today. Their explanation of the universe was based on tribal gods of earth, sky and water, with the Earth at the centre of the universe.

The **Babylonians** (c.2000–500BC) began regular observations of the stars and planets using much of the knowledge of the Sumerians, and we know that by 1600BC they had produced star catalogues and were able to predict planetary motions from their records of daily, monthly and yearly cycles of celestial motions. Their motivation for this was both agricultural and religious. The positions of the Sun and the Moon were used to devise calendars from which the time to plant crops could be determined but, like the Sumerians, the 'gods of nature' still formed the basis of their cosmology and world view.

Greek cosmology

Greek cosmology was based on mathematics – the Greeks applied geometry and number to develop models of the universe. The word 'cosmos' comes from the Greek word *kosmos* which linguistically has connotations with ideas of symmetry and harmony. Greek astronomy started with Thales of Miletus (b.624BC) who studied the ancient records of Mesopotamia and was able to predict a solar eclipse in 585BC. The Greeks believed that the distant stars were attached to a giant sphere whose centre was the centre of the Earth, and which rotated once a day.

The Greeks made an important step forward when they discovered that the Earth was a sphere. Aristotle (384–325BC) advanced arguments that this was so, by observing that the pole star *Polaris* remains stationary while all other stars revolve around it. Aristotle also noticed that some stars could be seen from Egypt but not from Greece, and the only way to explain these facts was to regard the Earth as spherical. Using simple trigonometry, Eratosthenes in about 270BC worked out the circumference of the Earth to be about 40 000 km which is remarkably close to the correct value. Aristarchus of Samos (310–230BC) actually proposed that the Earth went round the Sun in one year and in doing so proposed the first Sun-centred or **heliocentric** model for the cosmos, as opposed to a **geocentric** or Earth-centred model. Unfortunately Aristarchus's ideas never caught on, and Greek cosmology reverted to the geocentric model.

Ptolemy's model of the universe

We have already mentioned Hipparchus of Nicaea (see Chapter 1) who compiled detailed star catalogues and devised a stellar magnitude system, but it was Claudius Ptolemy who, from AD127 to 151, developed a systematic cosmological model that endured for over 1400 years. Ptolemy revised and extended Hipparchus's star catalogues and formulated a geocentric model of the universe based upon the properties of circles.

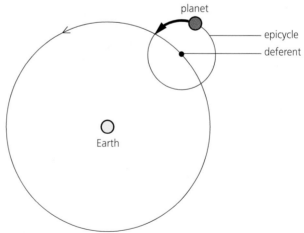

Figure 10.1 Ptolemy's geocentric system of planetary motion. A planet moves in a small circle or epicycle. The centre of the epicycle, called the deferent, moves around the Earth in a circle

Ptolemy was aware that the stars appear to remain fixed, but the Sun, Moon and planets change their positions and move between the constellations depending on the time of year. Particularly puzzling was the movement of some of the planets, which at times seemed to stop and move backwards in retrograde motion before resuming their normal paths.

Like all Greek philosophers, Ptolemy was imbued with a sense of 'form' or harmony but this retrograde motion of the planets did not fit the idea of a uniform circle which was considered to be a perfect form. Ptolemy devised a geocentric system based on a series of concentric circles with the Earth at the centre and the stars remaining fixed on an outermost celestial sphere. In order to explain the retrograde motion of the planets, a system of smaller circles called **epicycles** described the path of a planet around the Earth. The centre of the epicycle, called the **deferent**, also moved around the Earth in a circle (Figure 10.1). The **Ptolemaic model** was able to account for many of the observed motions of the planets, including the prediction of their positions to some accuracy, and this is one of the main reasons it lasted as long as it did despite its complicated use of the geometry of circles.

The Copernican revolution

Nicolaus Copernicus (1473–1543) was a Polish churchman who was greatly interested in astronomical theory. After studying the Ptolemaic model he came to the conclusion that it could be greatly simplified if the Sun was placed at the centre with all the planets revolving around it and only the Moon revolving around the Earth. This was the first step back to a heliocentric system since Aristarchus of Samos, and Copernicus knew that this theory would put him at odds with the teachings of the Roman Catholic Church, which at the time strongly maintained that the Earth was the centre of the universe.

Copernicus's theory was published in 1543 in a book called *De Revolutionibus Orbium* (Concerning the Revolutions of the Celestial Bodies) as he himself lay dying. Copernicus did not introduce the notion of any centripetal force between the Sun and the planets and he still believed that the planets went round the Sun in perfect circles. He even had to bring back epicycles, but his important contribution was to place the Sun at the *centre* of motion.

Tycho Brahe (1546–1601) was a Danish astronomer who kept careful observations of planetary positions. He did not accept Copernicus's heliocentric theory. When he died, he left his recorded observations to his assistant the German Johannes Kepler (whom we have already met in Chapter 2). Kepler believed in Copernicus's model and used Brahe's planetary observations to construct a heliocentric model based not on circles but on ellipses. In Chapter 2 we looked at the properties of an ellipse and how Kepler proposed a model in which the planets moved around the Sun in elliptical paths with the Sun occupying one of the foci and from this, Kepler formulated his three laws of planetary motion. Kepler found that his model predicted planetary positions to a far greater accuracy than the

Copernican and Ptolemaic systems, mainly because it did away with the tradition of form and relied on the precision of Brahe's measurements and geometrical analysis.

Galileo Galilei (1564–1642) was a firm believer in Copernicus's heliocentric model and was one of the first to use a telescope to make astronomical observations. Galileo found that the Moon had mountains and craters and that the Milky Way could be resolved into vast numbers of stars. He discovered that the planet Venus displayed phases similar to that of the Moon and that the planet Jupiter possessed four moons which changed position nightly as they revolved around it. This last observation was significant as it conclusively showed that there was a second centre of motion in the universe, completely contradicting the Ptolemaic idea that the Earth was at the centre of everything.

Galileo did not keep these discoveries quiet and was eventually forced by the Church in Rome to retract his belief that the Earth goes round the Sun. In 1632 Galileo published a book called *Dialogue on the Two Chief World Systems*, which compared and contrasted the Ptolemaic and Copernican systems and came out in favour of the Ptolemaic model but, in reality, it was a thinly disguised document supporting the heliocentric model of Copernicus. In the *Dialogue*, the cosmology of Galileo placed the Sun at the centre with the planets in circular orbits again – Galileo evidently took little notice of Kepler's work! Beyond the orbit of Saturn lay the spherical shell of stars envisaged by the Greeks, except that Galileo allowed for the possibility that the stars might extend to infinity which was a significant departure from the 'closed' universe of Greek thinking. It was Galileo who founded the science of mechanics and developed concepts of force, inertia and motion, and his experiments led him to conclude that all matter fell towards the Earth with the *same* acceleration.

However, none of Copernicus, Kepler or Galileo was able to demonstrate the physical cause that kept the planets in orbit around the Sun. It remained for Newton, with his Laws of Universal Gravitation and Laws of Motion, to supply the answer. Gravity is the key player in the universe and it is gravity acting between the Sun and the planets that is the underlying physical cause.

The cosmology of Newton demanded that the universe be infinite and Newton argued that if it wasn't then gravity would eventually pull all matter to one centre resulting in a single large mass. The fact that this does not happen must mean that there are an infinite number of masses in the form of stars that exert gravitational forces on each other, so that the overall effect is that there is no net force in any direction and the universe is prevented from contracting.

Copernicus, Kepler, Galileo and Newton effectively dealt the death blow to geocentric models of the cosmos. We will now look at how, by building on their achievements, the modern theory of the cosmos has developed, and how this has led to a new cosmology that can possibly explain the origin and fate of the universe.

A new theory of gravity

Despite the success of Newton's mechanics in explaining the motion of the planets, there were still some things that did not fit. One of these was the orbit of the planet Mercury. As Mercury orbits the Sun, the orbit itself rotates and this effect, called **precession**, can be measured (Figure 10.2). It was found that the degree of precession did not match that predicted by Newtonian mechanics and although the discrepancy was very small it was nonetheless real and outside the bounds of experimental error.

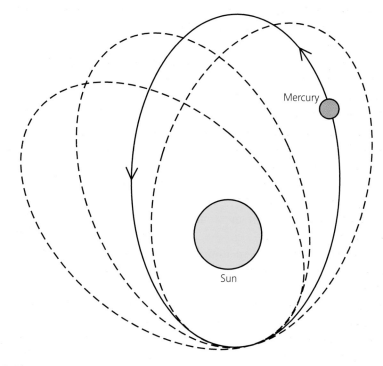

Figure 10.2 The precession of Mercury's orbit cannot be explained by Newtonian mechanics

The Michelson–Morley experiment

During the 19th century physicists believed that all electromagnetic radiation, including light, was a transverse wave motion that propagated through a luminiferous (light-carrying) substance or 'ether' which occupied all space. The idea of the ether came about because of the belief that all waves needed a medium through which to travel.

This concept of a universal ether was knocked on the head by two American physicists, Albert A. Michelson (1852–1931) and Edward W. Morley (1838–1923), who showed that the concept of a universal ether was fundamentally flawed. It was commonly supposed that the ether was motionless and that the Earth moved through it. In other words the ether was moving relative to the Earth. If this was the case, then light from a source travelling in the same direction as the Earth's motion ought to travel more rapidly than light travelling at right angles to it.

Michelson and Morley used an optical interferometer system to look for the interference fringes that would be produced by two beams of light having different path lengths due to their differing velocities. The surprising result was that Michelson and Morley could detect no such interference fringes whatever orientation their interferometer was placed in. The speed of light seemed to be constant in whatever direction it was travelling relative to an observer and they concluded that the ether did not exist.

The Special Theory of Relativity

Albert Einstein pointed out that without an ether, the notion of 'absolute rest' or 'absolute motion', which is an essential feature of Newtonian mechanics, would have to be abandoned. All motion was *relative* to a 'frame of reference' or platform from which measurements are made, usually chosen for the observer's convenience. Einstein made two key assumptions or **postulates** which he incorporated into his **Special Theory of Relativity**:

Postulate 1   *The laws of physics are the same in all inertial frames of reference.*

Postulate 2   *The speed of light in a vacuum is the same in all inertial frames of reference.*

By an **inertial** frame of reference we mean a reference frame that is *not* accelerating, i.e. it is at rest or moving with uniform velocity. The Special Theory of Relativity shows how measurements of length, time and mass are affected by motion; and by assuming the constancy of the speed of light for all observers, Einstein showed that some very peculiar effects take place.

Length contraction

A ruler of 'rest length' l_0 when travelling with a velocity v will 'shrink' to a length l given by:

$$l = l_0\sqrt{\left(1 - \frac{v^2}{c^2}\right)}$$

where c is the speed of light.

Time dilation

The rate at which time passes varies with the velocity of motion. This 'time dilation' is described by the equation:

$$t = \frac{t_0}{\sqrt{1 - \dfrac{v^2}{c^2}}}$$

where t_0 is a time interval measured on a clock which is at rest with respect to an observer and t is the time interval measured on a clock travelling with velocity v.

Relativistic mass

For momentum conservation to hold for higher speed collisions there must be a new definition of momentum. Instead of $p = m \times v$ we must have

$$p = \frac{m_0 v}{\sqrt{\left(1 - \dfrac{v^2}{c^2}\right)}}$$

where m_0 is called the **rest mass** of the object or its observed mass when at rest. This is normally interpreted as saying that the mass of a body increases with speed as

$$m = \frac{m_0}{\sqrt{\left(1 - \dfrac{v^2}{c^2}\right)}}$$

Another consequence of special relativity is that all notions of **simultaneity** have to be abandoned. Space and time are no longer separate independent quantities as in Newtonian mechanics but instead become linked together in a 'space–time continuum'.

These effects of length contraction, mass increases and time dilation with velocity are real but we do not normally observe them in everyday life since they only become noticeable at velocities approaching an appreciable fraction of the speed of light.

The term $\sqrt{1 - (v^2/c^2)}$ that appears in each of these equations explains why nothing can travel faster than light. If v approaches c then $\sqrt{1 - (v^2/c^2)}$ approaches zero and length becomes infinitely small, as do time intervals, and mass becomes infinitely large.

Box 10.1 The lifetime of muons – a test of special relativity

Muons (μ) are elementary particles that are created by cosmic rays high in the atmosphere. They can be positively or negatively charged and decay into an electron or positron and a neutrino and antineutrino by the reaction:

$$\mu^{\pm} \longrightarrow e^{\pm} + \nu + \bar{\nu}$$

Because they decay, muons are like radioactive clocks. With a charged particle detector we can detect the arrival of a muon and then at certain time later the production of an electron or positron, and so measure the decay time.

The lifetime of muons in a laboratory reference frame is found to be 1.5×10^{-6} seconds. However, the muons travelling down towards the surface of the Earth have an observed lifetime in our frame of about 14×10^{-6} seconds. The reason for this is relativistic **time dilation**.

The muons are travelling at speeds of about $0.994\,c$ and if we use the time dilation equation of special relativity

$$t = \frac{t_0}{\sqrt{1 - \frac{v^2}{c^2}}}$$

we find that from our point of view the muon lives for

$$\frac{1.5 \times 10^{-6}}{\sqrt{1 - (0.994)^2}} = 14 \times 10^{-6}\,s$$

To us, the muon's decay time seems to have been slowed by a factor of almost 10 so that it has enough time to reach the Earth before decaying.

From the muon's point of view it still has a lifetime of 1.5×10^{-6} s but the thickness of atmosphere that it has to travel through has been foreshortened by length contraction. Using

$$l = l_0 \sqrt{\left(1 - \frac{v^2}{c^2}\right)}$$

the muon finds the distance it has to travel has contracted by the same factor, so that it has a shorter distance to go before it reaches the ground.

Special relativity could explain Michelson and Morley's result and, in addition, Einstein was able to show the equivalence of mass and energy $\Delta E = \Delta mc^2$ which, as we have seen in Chapter 7, is of vital importance for the generation of energy in stars.

The General Theory of Relativity

In 1915 Einstein published a paper referred to today as the **General Theory of Relativity** which dealt with the more general case of bodies undergoing *acceleration* in a frame of reference as opposed to the Special Theory which only considered the motion of bodies with uniform velocity. Such a frame of reference in which bodies accelerate is called a **non-inertial** frame.

Einstein introduced a set of equations called the 'field equations' of general relativity which did away with the Newtonian concept of gravity as a force and, instead, relied on a *geometrical* description of space and time in which bodies move along paths called **geodesics** where space has been 'curved' by the presence of mass (in geometry a geodesic is the shortest distance between two points). The larger the mass, the greater the curvature and the deflection of the body's motion (Figure 10.3). The mathematical description of the field equations is beyond the scope of this book and in this section we just discuss the important concepts of general relativity.

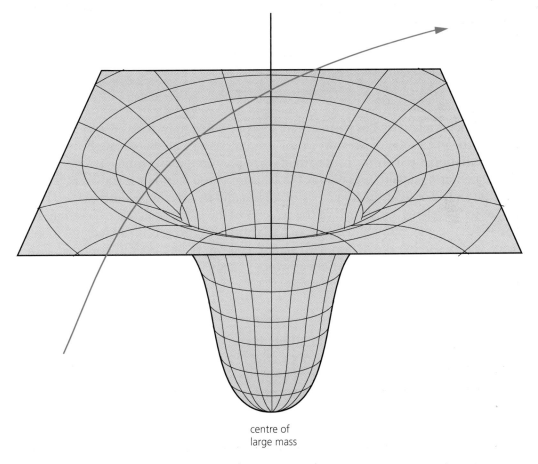

centre of
large mass

Figure 10.3 The curvature of space by a gravitational field. General relativity predicts that matter curves space–time. This diagram represents a two-dimensional analogy of the geometry of space–time near a massive object

Principle of equivalence

Einstein proposed the **Principle of Equivalence** which stated that *you cannot distinguish between a uniform gravitational field in a non-accelerating inertial reference frame and a uniformly accelerating (non-inertial) one.* This is best illustrated by example.

Suppose you were in a lift in which the cable had snapped and you were falling freely. You would experience exactly the same effects as an astronaut floating freely in interstellar space and would not be able to tell the difference if you were in an isolated chamber and had no way of knowing where you were. It appears that acceleration is able to 'cancel out' the effect of a gravitational field. The Equivalence Principle can be stated in another way. From Newton's 2nd Law, a mass m can be defined in relation to a force F acting on it and the acceleration a it experiences. Mass defined in this way is called **inertial mass** and we write it as $F = ma$. Alternatively, we can define mass using Newton's Law of Gravitation $F = GMm/r^2$ where r is the distance between two masses M and m. When defined like this, mass is called **gravitational mass**. The mass m that appears in the equation for inertial mass is, said Einstein, the same m that appears in the equation for gravitational mass – the two are equivalent even though they are defined differently! In recent years experiments have confirmed this to be true to at least 1 part in 10^{11}, which is an unprecedented level of agreement.

Gravitational red shift

So far general relativity has passed every test presented to it. It has been able to explain the precession of Mercury's orbit, and it has also predicted an effect called the **gravitational red shift**. In this situation, a photon moving from a strong to a weak gravitational field loses energy but because of the fact that it travels at the constant speed of light, its frequency decreases and its wavelength increases. This means its wavelength is shifted to the red end of the electromagnetic spectrum. In the 1960s the gravitational red shift of light from the Sun was measured and found to agree with Einstein's prediction with great precision. (*Don't* get this confused with the red shift due to the Doppler effect which is caused by the motion of an object receding from an observer – see Chapter 2.)

Bending of light

Another consequence of general relativity is that light should be deflected by a strong gravitational field. This was put to the test with the total solar eclipse of 19th March 1919. A number of bright stars were in the vicinity when the Sun was in totality and, if the General Theory was correct, their apparent positions would have altered slightly from those they occupied six months previously when they were observed in the night sky. This was because the light from the stars would have been bent as it passed close to the Sun. An expedition was sent to observe the eclipse and found that the altered positions were exactly in accordance with the General Theory. It is also the General Theory that predicts gravitational lensing and the extreme curvature of space–time around a black hole as we saw in Chapter 9.

Flat and curved space-time

Newtonian mechanics demands that the geometry of the universe is flat and that masses interact gravitationally by means of forces, with space and time being independent quantities. Einstein, however, said that the geometry of the universe can have any shape depending on the distribution of matter in it, and that space and time are linked together. Objects move along the curvature of space-time caused by the presence of mass. Newton's mechanics is not, however, invalidated as it turns out to be a just special case of the General Theory. To sum up Einstein's mechanics we repeat what we stated in Chapter 9: *'space–time tells matter how to move'* and *'matter tells space–time how to curve'*, and we will return to these ideas later in this chapter when we discuss the shape of the universe.

Olbers' paradox

Why is the sky dark at night? On the face of it this may seem a strange question to ask but consider this: if the universe were infinite and contained stars more or less homogeneously, then everywhere we look we would eventually intercept light coming from some star and so the sky would appear uniformly bright. This question was discussed in a paper written in 1823 by a German amateur astronomer Heinrich Wilhelm Olbers (1758–1840). His discussion is quite involved, but the essential feature of the argument is that if the universe is infinite in space then although the light received from each star decreases as $1/(\text{distance})^2$, i.e. as an *inverse square* law, the number of stars *increases* as the $(\text{distance})^2$, a *square* law. The two scaling factors cancel out and we conclude that the sky should appear uniformly bright. But this is not what we observe. How can we resolve this paradox?

The answer lies in the fact that the speed of light is *finite*. We view the universe at some stage in its past because the light from that era has taken a certain time to reach us. Olbers' paradox is powerful evidence that the universe at some point must have had a beginning, since if it was infinitely old, then the sky would be bright. Later on we will see how cosmologists can estimate just how old the universe is. More significantly, Olbers' paradox can be explained if we consider the universe to be expanding.

The expanding universe

The Cosmological Principle

The universe contains structure at every scale from the nuclei of atoms up to the large-scale clustering of galaxies. In the modern view of the universe, cosmologists make two key assumptions about what we can see when we look out into space.

The first is that the universe is more or less **homogeneous** at scales greater than a few hundred Mpc. This means that if, for example, you took a cube of space anywhere in the universe say 500 Mpc square, then you would expect to find the *number* of galaxies it contained to be more or less the *same* as in a cube of the same

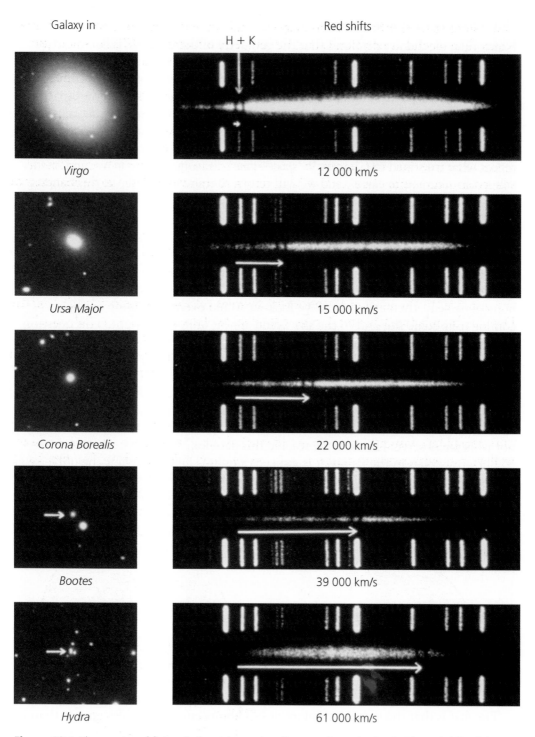

Figure 10.4 The spectra of five galaxies at increasing distances from the Earth. The red shift of the H and K lines of ionised calcium is shown, and the recessional velocities, calculated using the Doppler formula. The more distant the galaxy, the greater the red shift

dimensions at some other location in space – the universe appears 'smooth' at these scales. The second assumption is that the universe is **isotropic**. What cosmologists mean by this is that it looks the *same in every direction*. While there may be local variations in the exact arrangements of galaxies and galaxy clusters, all observers located anywhere in space would see the same large-scale structure of the universe that we observe from the Earth.

These two assumptions of homogeneity and isotropy taken together are called the **Cosmological Principle**. As far as the limit of our observations allows, this principle seems to be true, but if, for example, astronomers were to discover a huge uniquely shaped structure some thousands of Mpc across, then it would have to be abandoned!

The Hubble Law

During the 1920s an American astronomer Edwin Hubble (1889–1953) made a very important discovery. Hubble, together with others, had noticed that the spectral lines of distant galaxies were red-shifted indicating that they were moving away from us (Figure 10.4). Using standard candle methods as distance indicators, he established the distances to a number of them for which radial velocities were measured using the Doppler formula (see Chapter 2, page 35). Hubble plotted a graph of radial velocity versus distance and obtained a straight line, which showed that the radial velocity was directly proportional to the distance. The result is known as the **Hubble Law** and if v is measured in km s^{-1} and distance r in Mpc we get the relation:

$$v = H_0 \times r$$

where H_0 is known as the **Hubble constant** with a unit of km s^{-1} per Mpc. The value of the Hubble 'constant' is the subject of much debate. Its accuracy depends on how well we can measure the distances to galaxies by independent methods (see Box 10.2). The current value is thought to lie between the 50 and 100 km s^{-1} per Mpc and in this book we will adopt the currently accepted value of $H_0 = 75$ km s^{-1} per Mpc.

Given the radial velocity of an object, the Hubble Law becomes a method of measuring distance.

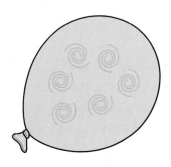 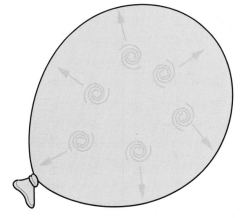

Figure 10.5 Galaxies marked on a balloon get further apart as the balloon is inflated. The surface of the balloon is a two-dimensional analogy of the expansion of the universe

WORKED EXAMPLE 10.1

A galaxy in the *Virgo Cluster* is observed to have a radial velocity of 1200 km s^{-1}. How far away from us is it?

Rearranging the Hubble Law we get:

$$r = \frac{v}{H_0} = \frac{1200 \text{ km s}^{-1}}{75 \text{ km s}^{-1} \text{ per Mpc}} = 16 \text{ Mpc}$$

The Tully–Fisher relation

A more recent method of measuring cosmological distances is called the **Tully–Fisher relation** named after Brent Tully and Richard Fisher, the two American astronomers who discovered it.

Every galaxy rotates. We mentioned in Chapter 1 that the Sun, together with other stars, rotates around the centre of the Milky Way. In the 1970s, astronomers found that there was a proportional relationship between the rotational speeds of *spiral* galaxies whose distances were a few tens of Mpc, and their luminosities. From our discussion of circular motion and angular momentum (Box 2.2, page 43 and Box 9.3, page 170), the rotational speed of an object is related to its total mass. The

Box 10.2 Measuring cosmological distances

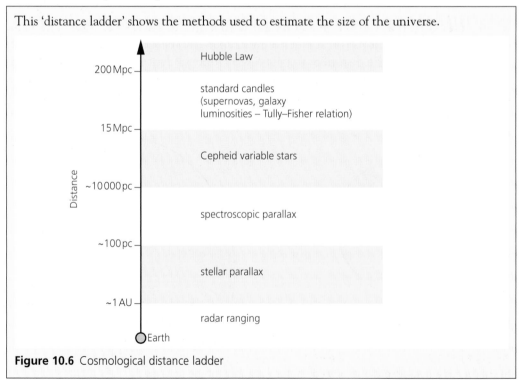

Figure 10.6 Cosmological distance ladder

rotational speeds of spiral galaxies are a measure of the galactic mass, and this in turn is related to the total luminosity. The faster they rotate, then the brighter they are and the Tully–Fisher relation allows us to obtain an estimate of a galaxy's luminosity by observing how fast it rotates. The rotational velocity is determined by observing either the Doppler broadening in the galaxy's spectrum at visible wavelengths, or at the 21 cm radio emission from the HI regions in its interstellar gas lanes (see Box 6.1, page 117). Once an estimate of the luminosity (brightness) is made, then we can compare it with the measured observed brightness using photometry, and determine the distance to the galaxy using Pogson's Law. The Tully–Fisher relation can be calibrated and applied to more remote spiral galaxies.

Astronomers are cautious about using the Tully–Fisher relation because the number of calibration galaxies is presently quite small, but it does, nonetheless, provide another method of climbing the cosmological distance ladder.

The distances we measure using the Hubble Law depend on how well we know the value of H_0. Astronomers can measure the distances of galaxies by independent methods up to a distance of about 400 Mpc. Beyond this we cannot be sure that the Hubble Law remains true; in fact we will see later that there are reasons to think that it departs from a straight line at very large distances, and astronomers devote much experimental effort in determining an accurate value for the Hubble constant.

To understand what an expanding universe means, cosmologists often use an analogy of inflating a balloon. Imagine a balloon with galaxies marked on it being blown up (Figure 10.5). As the balloon inflates the galaxies get further away from each other. If an ant were sitting on one of the galaxies then it would see all the others receding from it, in whatever direction it looked. The galaxies closer to it would be receding more slowly, and the ones further away more rapidly.

The galaxy the ant is on is not the 'centre' of the balloon as the balloon's surface has no centre. The surface of the balloon has no edge, either. The ant could walk all over the surface of the balloon without ever finding an edge and return to where it started from. This is what the Hubble Law tells us. The universe has no centre and no edge. The most distant galaxies are moving away from us at a faster rate than the nearest ones. Space itself is expanding and it is as meaningless to talk about what it is expanding into as to ask 'what is the edge of a sphere?'

The cosmological red shift

If the universe is expanding then this accounts for why light from distant galaxies is red-shifted. As photons travel through space their wavelengths become 'stretched' and therefore longer than usual. The further a photon has to travel to reach us the more its wavelength will have been stretched out and the greater its red shift is measured to be. Cosmologists distinguish between a red shift caused by the Doppler effect and one that is caused by the expansion of the universe. Although the Doppler formula is used in both cases to calculate velocities, it is important to understand that a Doppler shift is caused by a body's *motion through space* whereas a **cosmological red shift** is caused by the *expansion of space*. Box 10.2 shows the 'distance ladder' used in measuring the size of the universe.

The Big Bang

The age of the universe

If the universe has been expanding for some while then it is reasonable to assume that at some time in the past, all the matter it contains must have been closer together and concentrated in a state of high density. For a galaxy moving with constant velocity v the distance D travelled in time t is:

$$D = v \times t$$

and the travel time is:

$$t = \frac{D}{v}$$

But from the Hubble Law we know that:

$$v = H_0 \times D$$

so that

$$t = \frac{D}{H_0 \times D} = \frac{1}{H_0}$$

If we use the value for H_0 of 75 km s^{-1} Mpc, remembering that 1 Mpc $= 3.1 \times 10^{19}$ km, we get:

$$t = 1 \times \frac{3.1 \times 10^{19} \text{ km}}{75 \text{ km s}^{-1}} = 4.13 \times 10^{17} \text{ s}$$

Since there are 3.15×10^7 s in one year this estimate is about 1.3×10^{10} years. Therefore this puts the time when the universe started to expand as some 13 000 million years ago. From this, we can see that the reciprocal of the Hubble constant tells the age of the universe and *its value changes with time as the universe expands*. The Hubble 'constant' is therefore more accurately called the **Hubble parameter** and cosmologists are very interested in obtaining a reliable estimate of it.

The expansion of the universe suggests that at this time in the past, all the galaxies were together in one place. In fact everything, including radiation and matter, was concentrated at a single point. The importance of the Hubble Law is that it implies that the universe is *finite* and this is what we would expect if Olbers' paradox is to be resolved. We cannot see any light from galaxies more than 13 000 million light years away and this marks the limit of the *observable* universe. If the universe were infinite then light from galaxies beyond this limit wouldn't have had enough time to reach us yet.

What was the Big Bang?

If all the matter in the universe was at some time concentrated at a single point then about 13 000 million years ago some kind of gigantic explosion must have occurred to cause the universe to start expanding. Cosmologists call this event the **Big Bang**. You might wonder at this point if the Hubble Law violates the Cosmological Principle, since it seems to pick out a unique point in space from where the universe began and a unique 'edge' to all the expanding matter. This in fact is not the case. The Big Bang did not happen at some point in space in an otherwise empty universe. The Big Bang is *space itself expanding* and the galaxies are moving with it just like the galaxies on the inflated balloon. In this way the universe cannot be said to have an edge and the homogeneity and isotropy required by the Cosmological Principle remains intact. The Big Bang started as a **singularity** in space–time where the universe had zero size and infinite density and temperature. Later in this chapter we will see how far astrophysics can take us in determining the physical conditions and the origin of matter soon after the Big Bang.

Evidence for a Big Bang

As well as the Hubble Law there is another important piece of evidence that points to a hot Big Bang as being the origin of the universe. In 1964 Arno Penzias (1933–) and Robert Wilson (1936–), two scientists at Bell Telephone Laboratories in New Jersey, were carrying out experiments using a microwave antenna for satellite communications. As they pointed the antenna towards the sky, their receiver registered a faint 'hiss' coming from all directions that would not go away. The hiss was highly isotropic, constant with time and could be detected at any time of day or year. What Penzias and Wilson had detected was a relic from the Big Bang – the thermal radiation from the Big Bang itself!

Cosmologists have theorised that shortly after the Big Bang occurred, the universe was filled with very high energy blackbody radiation consisting of gamma rays of very short wavelength. As the universe expanded it cooled, and the wavelength of this cosmological radiation red-shifted down to the microwave region of the electromagnetic spectrum with a blackbody temperature of about 3 K. This is in close agreement to that measured by Penzias and Wilson. In 1989 a space astronomy satellite called the **Cosmic Background Explorer (COBE)** was launched which measured the distribution of microwave radiation in more detail and confirmed a blackbody curve with a peak wavelength corresponding to a temperature of about 2.7 K.

The **cosmic microwave background** represents photons which started their journey when the universe was only 100 000 years or so old and in this way shows us something of the nature of the universe when it was very young. To understand what it was like before this, we will have to examine the physics of the early universe, which we will do shortly.

The future of the universe

Will the universe go on expanding forever? To answer this question depends on knowing the average density of matter it contains. As the universe expands, gravity will continuously act to slow down and decelerate the expansion. If the gravitational strength generated by this matter is too weak, then the universe will go on expanding and the galaxies will continue to move away from each other. In this case cosmologists say that the universe is **unbounded**.

If, on the other hand, the average density of matter is high enough, then gravity will halt the expansion and the universe will at some time in the future reach a maximum size. Gravity will then start pulling all the galaxies back towards each other and the universe will contract. This is analogous to deflating the balloon (Figure 10.5), and in this situation the universe is said to be **bounded**.

Critical density

Between an unbounded and bounded universe is the case where the average density of matter is just high enough so that the galaxies come to a halt in an infinite time when they are infinitely far apart, and cosmologists call this value the **critical density**.

We can work out a value for the critical density by using the escape speed formula that we derived in Chapter 2, page 44. We do this by modelling the expansion of the universe in a similar way to that of a mass escaping from the Earth. A rocket leaving the Earth with less than the escape speed will slow down and fall back, but one that is travelling with a speed *equal* to the escape speed will stop but only when it is an infinite distance away after an infinite time.

The escape speed v_{esc} from a sphere of mass M and radius R is given by:

$$v_{esc} = \sqrt{\frac{2GM}{R}}$$

and since mass = density $\times$ volume, $M = \frac{4}{3}\pi R^3 \rho$ and substituting this in the expression for v_{esc} we get

$$v_{esc} = \sqrt{\frac{2G(\frac{4}{3}\pi R^3 \rho)}{R}} = \left(\frac{8\pi}{3} G\rho\right)^{\frac{1}{2}} \times R$$

where ρ is the average density of matter in the universe. The Hubble Law says that $v = H_0 \times D$ and in this case we have $v = v_{esc}$ and $D = R$ the radius of the universe, and so we can write:

$$v_{esc} = H_0 \times R = \left(\frac{8\pi}{3} G\rho_c\right)^{\frac{1}{2}} \times R$$

where ρ_c is the critical density for escape velocity to be achieved. So by comparing terms

$$H_0 = \left(\frac{8\pi}{3} G\rho_c\right)^{\frac{1}{2}}$$

or

$$\rho_c = \frac{3H_0^2}{8\pi G}$$

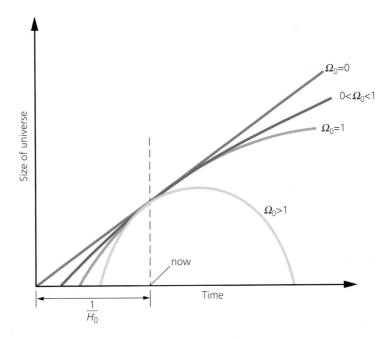

Figure 10.7 The density parameter Ω_0 determines the fate of the universe

Using the value of $H_0 = 75$ km s^{-1} per Mpc, $\rho_c = 1.1 \times 10^{-26}$ kg m^{-3}. If the average density of the universe is less than this value then the universe will go on expanding for ever. If it is greater then the expansion will eventually stop and the universe will collapse. If it is exactly equal to ρ_c, then the expansion will eventually stop at an infinite time in the future.

Cosmologists express these three scenarios by denoting the amount of deceleration by a **density parameter** Ω_0 which is defined as

$$\Omega_0 = \frac{\text{density of matter in the universe}}{\rho_c}$$

If the universe contained no matter at all then $\Omega_0 = 0$ and the universe would expand forever, and there would be no deceleration since there is no gravity to slow the expansion. For $0 < \Omega_0 < 1$ the universe is unbounded and will expand for ever. If $\Omega_0 = 1$ then this is the case when the expansion is exactly equal to the 'escape speed' and the universe will come to a halt at the end of time. For $\Omega_0 > 1$ the universe is bounded and will at some finite time collapse. The density parameter is illustrated in Figure 10.7.

Measuring the density of the universe

Strictly speaking, because of the equivalence of mass and energy, cosmologists talk about the 'mass–energy' density of the universe. But can we know what the average density of matter in the universe is? The value for ρ_c above is a calculated value and depends on how well we know the value of the Hubble constant; the value measured

201

by observational methods depends on how far we can see into space (using instruments like the Hubble Space Telescope) and at what wavelengths we can reliably detect radiation from all kinds of celestial objects.

Astronomers estimate the masses of galaxies and galaxy clusters in a large volume of space and obtain the average density by dividing the total mass by the total volume. This gives an average density of about 4×10^{-28} kg m^{-3}, suggesting that the universe will expand forever. However, estimates like this do not take into account any hidden material or 'dark matter' that cannot be detected by usual observational methods. Later we will see that some cosmological models require Ω_0 to equal unity which, if true, means that there is a considerable amount of missing mass that hasn't been discovered yet.

The shape of the universe

We have seen how Einstein's General Theory of Relativity says that 'matter tells space–time how to curve'. The degree of curvature depends on the amount of matter and so the overall geometry or shape of the universe should therefore depend on the distribution of matter across space.

Suppose we traced the paths of two galaxies formed after the Big Bang that start off parallel to each other as the universe expands. If we could observe their motion over a vast period of time then there are three possible courses they could take.

1 We might find that they remained parallel, staying the same distance apart over billions of light years. In this case space would be flat like a plane and the universe would have **zero curvature**.

2 A second possibility would be that we see the galaxies get closer to one another as the expansion of the universe increases and eventually converge at some point far from the Earth. The paths of the galaxies would be like the lines of longitude on a sphere which are parallel at the equator but cross over at the poles. The shape of the universe would be spherical and mathematicians would say that the universe has **positive curvature**.

3 The third possibility would be that we would see the galaxies become further apart as the universe expanded. The geometry of the universe would therefore be opposite to that of a sphere and such a shape is called **hyperbolic** with **negative curvature**.

Zero curvature corresponds to $\Omega_0 = 1$ with $\rho = \rho_c$ and the shape of the universe is like a flat plane. Positive curvature corresponds to $\rho > \rho_c$, meaning that the universe, like a sphere, has a closed shape (since a sphere is a closed surface). Negative curvature corresponds to $0 < \Omega_0 < 1$ with $\rho < \rho_c$ and the shape of the universe is hyperbolic. Which shape does our universe have? One answer to this question would be to see if the light from distant galaxies is bent by the curvature of space. If the universe is curved, a galaxy a long way away would appear larger than it actually is, just as an object appears bigger under a magnifying glass. This effect is not generally observed and, as we shall see in the next section, there are compelling theoretical reasons to believe that the universe is close to, or is in fact, flat.

The physics of the early universe

Where did all the matter in the universe come from? To answer this question cosmologists have turned to the sub-microscopic world of **particle physics**. Particle physics is the study of matter and energy at its most fundamental level and particle physicists seek to understand the basic structure of matter and the fundamental forces or **interactions** existing between matter in the universe. By doing this they hope to learn more about the various physical relationships that connect different kinds of particles together and how they can combine to make more complex forms of matter. Particles that cannot be broken down into smaller units are called **elementary** which is why particle physics is also known as elementary particle physics; an electron is an example of an elementary particle.

Particle physicists have discovered many different kinds of particles. The protons and neutrons in atomic nucleus are now understood to be composed of elementary particles called **quarks** in different combinations. In addition, matter is found to exist in two forms, matter and **antimatter**, which has the same mass but opposite electric charge.

Interactions between matter

In the universe, there are four basic interactions between matter:
1 **The gravitational interaction**. Gravity attracts all matter but is too weak to have significant effects at sub-atomic scales.
2 **The electromagnetic interaction**. This plays an important role in the forces between sub-atomic particles and is responsible for the phenomenon of electromagnetism.
3 **The strong interaction**. This is the force that binds atomic nuclei together and stops them flying apart due to the mutual repulsion of their protons.
4 **The weak interaction**. The weak interaction is a force that is involved in nuclear beta decay and other radioactive processes.

All these four interactions, or forces, are mediated by **carrier particles** which are elementary particles that 'carry' the force from one particle to another. Table 10.1 shows the properties of the basic interactions.

Table 10.1 Properties of interactions

Interaction	Strength (relative to strong)	Range/m	Carrier particle
Gravitational	10^{-38}	∞	Graviton
Strong	1	10^{-15}	Gluon
Electromagnetic	10^{-2}	∞	Photon
Weak	10^{-6}	10^{-18}	$W^{\pm}$ and Z

The W and Z particles were discovered in 1984 using the Super Proton Synchrotron particle accelerator at CERN in Geneva. The gluon has been inferred by indirect methods but the graviton awaits a quantum theory of gravity and has so far not been observed.

It is not the purpose of this section to give a detailed account of the fundamental nature of matter, but to show how some important ideas from particle physics can be used to construct a model that explains the origins of matter and accounts for the differing properties of the fundamental interactions. The science of particle physics is a stunning achievement of the theory of quantum mechanics, although there are still areas where it is incomplete, notably in describing gravity using quantum theory.

Pair production

Earlier we mentioned that cosmologists refer to the average *mass–energy* of the universe due to the equivalence of energy and matter through $\Delta E = \Delta mc^2$. The average density of matter in the universe today is estimated to be about 10^{-26} kg m^{-3}. Radiation is produced by stars and galaxies in a concentrated form but these only occupy a small fraction of the entire universe. By far the most dominant source is the cosmic microwave background which has a mass-density equivalent of about 5×10^{-31} kg m^{-3}, some 100 000 times smaller than the density due to matter alone.

Figure 10.8 The creation of a particle–antiparticle pair. The forked track on the left shows the creation of an electron and a positron from a gamma ray (not visible). The track on the right is a stray electron. The picture was taken inside a bubble chamber, which makes the paths of charged particles visible as a row of tiny bubbles

This suggests that, at the present moment in time, we live in a universe that is dominated by matter. However, cosmologists believe that in the universe's early history it was *radiation* that dominated and in order to understand why, we need to look at a key concept in particle physics called **pair production**.

Particle physicists have found that matter is created when two high-energy photons collide with each other. The photons disappear and all their kinetic energy goes into the production of a **particle–antiparticle pair**. This pair-production is a consequence of mass–energy equivalence; a picture of particles being created in this way is shown in Figure 10.8.

We can write pair production symbolically as:

$$\text{photon} + \text{photon} \longrightarrow \text{particle} + \text{antiparticle}$$

The reverse process is also possible. A particle and antiparticle can collide and annihilate each other producing two high-energy gamma ray photons:

$$\text{particle} + \text{antiparticle} \longrightarrow \text{photon} + \text{photon}$$

For pair production to occur two conditions must be satisfied.

1 The energy of the photons must be greater than the mass–energy of the created particles.
2 Pair production must obey the laws of conservation of energy and momentum.

Cosmologists model the early universe as a hot, high-energy photon gas and the energy the photons have depends on their temperature. Creation and annihilation of particles occurs in equal numbers so that as many particles are made per second as are being destroyed and the universe is in a state of **thermal equilibrium**. Particle physicists call this state of balance between matter and antimatter a **symmetry**.

WORKED EXAMPLE 10.2

Given that the average energy E of photons in a gas can be approximated by the equation $E = \frac{3}{2}kT$ (where k is Boltzmann's constant) at what temperature T must the gas be for the combined energy of two photons to create a particle–antiparticle pair consisting of an electron and a positron?

The equivalent energy of an electron at rest is $m_e c^2 = 9.1 \times 10^{-31}$ kg $\times (3 \times 10^8 \text{ m s}^{-1})^2 = 8.2 \times 10^{-14}$ J (or 0.51 MeV) and since a positron is created as well, the total energy required is 2×0.51 MeV $= 1.02$ MeV or 1.6×10^{-13} J. Using $E = \frac{3}{2}kT$ the temperature is

$$T = \frac{2E}{3k} = \frac{2 \times 1.6 \times 10^{-13} \text{ J}}{3 \times 1.4 \times 10^{-23} \text{ J/K}} \approx 10^{10} \text{ K}$$

Protons and antiprotons would need a higher temperature to create them because of their *higher mass*.

It is important to understand that the creation of different kinds of particles by pair production is dependent on the *temperature* of the photons.

The history of the universe divided into four eras

The Big Bang model requires that the universe started from a primordial fireball of infinitely high temperature and density. For reasons explained later, we will outline the development of the universe when its temperature was 10^{32} K and its age was greater than 10^{-43} s. We model the universe initially as a high-energy photon gas which, like any gas, cools when it expands. Cosmologists divide the history of the universe into four periods or **eras** of time, each corresponding to a particular range of temperatures. These are:

1 a **heavy particle** era   3 a **radiation** era
2 a **light particle** era   4 a **matter** era

Heavy particle era: temperature of universe $<10^{33}$ K, time after Big Bang $>10^{-43}$ s

During this period, the universe is hot enough for all massive elementary particles to be created by pair production. The universe is in thermal equilibrium and expands rapidly, with the first stable protons being formed at about 10^{-6} s.

Light particle era: temperature of universe $<10^{12}$ K, time after Big Bang $>10^{-4}$ s

The temperature of the universe is no longer hot enough for massive particles to be made. Lighter particles such as electrons and positrons can still be produced, and protons and electrons combine to make neutrons. We mentioned earlier that we live now in a matter-dominated universe and for this to be the case, there must have been an excess of particles over antiparticles during the transition from the heavy to light particle eras. Particle physicists refer to this condition as a **symmetry breaking**.

Radiation era: temperature of universe $<10^{10}$ K, time after Big Bang >10 s

In this era, neutrons and protons left over from the heavy and light eras interact to form the first stable nuclei. The most important nucleus to form is that of deuterium ^2_1H and from this, stable nuclei of helium ^4_2He and light helium ^3_2He as well as beryllium ^7_4Be and lithium ^7_3Li are manufactured by fusion reactions:

$$^2_1\text{H} + {}^2_1\text{H} \longrightarrow {}^4_2\text{He} + \gamma$$
$$^3_2\text{He} + {}^2_1\text{H} \longrightarrow {}^4_2\text{He} + {}^1_0\text{n} + {}^0_1\text{e}$$
$$^4_2\text{He} + {}^3_2\text{He} \longrightarrow {}^7_4\text{Be} + \gamma$$
$$^7_4\text{Be} + {}^0_{-1}\text{e} \longrightarrow {}^7_3\text{Li} + \upsilon$$

The manufacture of light elements from neutrons and protons formed in the heavy and light era is called **primordial nucleosynthesis** (don't get this confused with *stellar* nucleosynthesis which is the formation of heavier elements in the interior of stars that we discussed in Chapter 8. About 25% to 30% of helium is formed by mass, the rest being mainly hydrogen with trace amounts of beryllium and lithium. The Big Bang model predicts that the universe should contain a helium abundance of at least these percentages with more being created by nuclear reactions in stars, and observations of the chemical composition of various celestial objects supports this.

Matter era: temperature of universe <3000 K, time after Big Bang >10^6 years

In previous eras matter and radiation have still interacted with each other and have been locked together. The radiation could not escape and the universe was **opaque** to radiation. However, when the temperature of the universe has dropped to 3000 K, the first hydrogen and helium atoms form and matter and radiation are no longer coupled together. The universe becomes **transparent** to radiation and cosmologists call this event **decoupling**. The decoupling of radiation and matter marks the stage at which the radiation expands with the universe, leaving the matter to interact with itself via gravity.

Due to local variations in density, the matter starts to clump together and material condensation occurs, leading cosmologists to believe that it was from the end of this era that large-scale structures such as galaxies could first start to form. The radiation that spread out after decoupling now comes to us as the 3 K cosmic microwave background which, because of its smoothness, suggests that before this event matter and radiation must have been very uniformly distributed. Exactly how galaxies formed after decoupling is far from clear, and it is hoped that instruments such as the HST will enable cosmologists to peer further back in time so that we can see galaxies in the early stages of their formation.

The origin of the four basic interactions

In our discussion of the origin of matter we have said nothing about where the forces of nature, summarised in Table 10.1, come from. Particle physicists believe that despite their different characteristics, if particles were to collide at very high energies these forces would be indistinguishable from each other and have the same strength. In other words, they would all be **unified** into a single 'superforce' and theories which attempt to unify the four interactions in this way are called **Grand Unified Theories** or **GUTs** for short.

In the 1970s particle physicists were able to show theoretically that the weak and electromagnetic forces, collectively called the **electroweak force**, could be unified at energies of about 100 GeV corresponding to temperatures of 10^{15} K. The unification of the electroweak and strong force does not occur until about 10^{14} GeV and, according to GUTs, all forces are unified at 10^{19} GeV. Particle accelerators have been built that can operate at energies of a few hundred GeV, but it is simply not possible to construct them capable of producing the full range of unification energies.

The Big Bang model, though, offers a way of testing GUTs. According to GUTs, during the first 10^{-43} s of the universe's existence, when it was at a temperature exceeding 10^{32} K, all the forces were unified as a single force. As the universe cooled, the forces of nature 'froze out' to the four interactions that we know today (Figure 10.9). This would account for the balance of matter over antimatter; before the strong force froze out, equal numbers of particles and antiparticles were being created and destroyed, but after the strong force froze out this symmetry was broken, leading to an excess of matter.

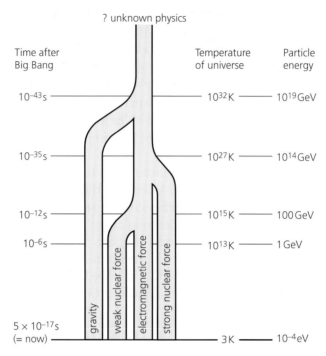

Figure 10.9 Grand Unified Theories. At temperatures greater than 10^{32} K during the first 10^{-43} s after the Big Bang, the forces of nature we know today were unified as a single force

GUTs also predict that neutrinos created in the Big Bang should have a small mass. If this is so then they would be a candidate for dark matter and the value of the mass–energy density might be altered so much that $\rho > \rho_c$, making the universe bounded; but alternatively, they might supply just enough missing mass to make $\rho = \rho_c$. Cosmologists believe that after decoupling, neutrinos may have aided galaxy formation within the accepted age of the universe. Also, astrophysicists think that neutrinos with mass might help solve the solar neutrino problem.

Problems with the Big Bang model

Despite its success, the standard Big Bang model still leaves some unanswered questions. The first is to do with the geometry of space and time. Observations suggest that Ω_0 is very close to 1, suggesting that we live in a flat universe. What special conditions in the Big Bang would have led to the mass–energy density being exactly equal to the critical density ρ_c?

It turns out that if the average density of matter during the Big Bang even slightly deviated from ρ_c then this variation would have rapidly multiplied as the universe expanded leading to a value very far from $\Omega_0 = 1$. The fact that Ω_0 is close to 1 now, means that some mechanism in the Big Bang ensured the near-flat universe we see today. This is called the **flatness problem** and it is explained by introducing a concept called **inflation**.

Cosmologists think that when the universe was between 10^{-43} and 10^{-30} s old, it expanded extremely rapidly. This had the effect of moving much of the material near where we are today very much further away. In the same sense that we cannot see the curvature of the Earth from the scale of a school playing field, so the universe subsequently appears flat. Inflation also explains why the cosmic microwave background is isotropic.

The **horizon problem** poses the question of how different parts of the universe could be at the same temperature when they are further apart by more than their light horizon (or the distance that could be travelled at the speed of light in a given time). When we observe the microwave background, we are looking at radiation from parts of the universe that were once very close together, but are now so much further apart that it seems the photons must have travelled faster than light!

The answer is that, before inflation, the universe was small enough for all parts to reach the same temperature within the light horizon. After inflation, this uniformity of temperature is maintained at larger distances even though it now takes considerably longer for light to travel across the universe. Note that inflation does *not* violate special relativity's postulate that the speed of light is constant since it is the expansion *of* space and not expansion *through* space that is at work here.

Before 10^{-43} s: the Planck time

In our brief explanation of the physics of the early universe we have discussed its evolution from 10^{-43} s, but what was the universe like before this? Indeed you might ask what came before the Big Bang? Earlier in the chapter we saw that general relativity regards space and time as linked together in a smooth continuum, but at very small scales of length and time, quantum mechanics predicts that this representation of space–time breaks down. This is because of a law in quantum mechanics called the **Heisenberg Uncertainty Principle** named after the German physicist Werner Heisenberg (1901–76).

The Uncertainty Principle tells us that there always exists a degree of uncertainty between the position of a particle and its momentum. The more accurately you measure the speed of a particle the less certain you are of its position – the Uncertainty Principle makes it impossible to measure any particle's exact position and exact momentum simultaneously. The principle also extends to energy and time. As the time available to make an energy measurement becomes smaller, the degree of uncertainty in the amount of energy becomes greater. Mathematically the uncertainty relations can be expressed as:

$$\Delta x \times \Delta p \geqslant h \qquad \text{and} \qquad \Delta E \times \Delta t \geqslant h$$

where Δx, Δp, ΔE and Δt are the uncertainties in position, momentum, energy and time, and h is Planck's constant. Accuracy in one of these measurements is at the expense of its corresponding partner. It's rather like trying to measure, say, the position of a finely balanced top spinning on a table with a metre rule. As soon as the rule touches the top it spins away in an unpredictable direction and the certainty of its original position is lost since the act of measurement has disturbed the system.

By using the Uncertainty Principle, cosmologists are able to arrive at an interval after which space and time become measurable, called the **Planck time** t_p, which is made up of the fundamental constants G, h and c:

$$t_p = \sqrt{\frac{Gh}{c^5}} = \sqrt{\frac{(6.67 \times 10^{-11} \text{ N m}^2 \text{ kg}^{-2}) \times (6.63 \times 10^{-34} \text{ J s})}{(3.00 \times 10^8 \text{ m s}^{-1})^5}} = 1.35 \times 10^{-43} \text{ s}$$

From the beginning of the Big Bang at $t = 0$ to $t = t_p$, we do not know how the universe behaved and we have no theory that can explain what happened using quantum mechanics or general relativity. Time itself began at the Big Bang and to say what came 'before' is a meaningless statement as time didn't exist then!

Particle physicists believe that the vacuum of space is not empty but is seething with pairs of particles and antiparticles that are constantly being created and destroyed. These particles exist for so brief a time (less than 10^{-21} s) that they cannot be directly observed and are called **virtual particles**. In fact the very act of observation would make them real. Energy is 'borrowed' from the vacuum to create them and repaid almost instantly. In order to conserve charge they are created in pairs of matter and antimatter, and annihilate themselves continuosly and at any instant; the vacuum is full of such virtual pairs. These particles may only exist for a time that satisfies the Uncertainty Principle, but as the universe rapidly expanded they became separated and appeared as real particles after the universe was older than the Planck time. Earlier, we mentioned that the Big Bang began from a **singularity** and we have already mentioned singularities in connection with black holes (Chapter 9). The gravitational energy associated with the Big Bang singularity provided enough energy for these particles to be created in vast quantities and fill the universe as it expanded.

The main point about all these concepts is that the Planck time sets a limit to what we can know about the very early universe according to our current knowledge of physics; however, theoretical particle physicists are currently working on a quantum theory of gravity and progress continues.

Echoes from the creation

In our discussion of cosmology one of the most remarkable things to emerge is the union of astrophysics with particle physics to help explain the origin of the universe. Understanding the very small helps us to understand the very big and we will end this chapter with a look at an exciting result from the Cosmic Background Explorer (COBE) satellite. Figure 10.10 shows an image of the microwave background taken from COBE. The image is displayed as a temperature map showing the distribution of microwave energy across the sky.

The high degree of uniformity of the microwave background means that the early universe must have been smooth and homogeneous but soon after the decoupling of radiation from matter, density fluctuations in this otherwise even distribution must have occurred from which galaxies and other large-scale structures formed. Careful image processing of COBE images such as this one show that these fluctuations have

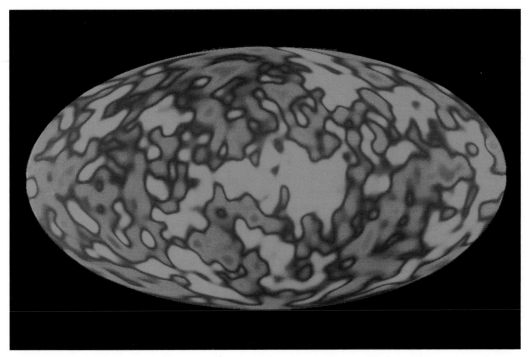

Figure 10.10 Temperature fluctuations in the microwave background. The variations are very small, of the order of 10^{-4} K

indeed been detected although the variations in temperature are only a few millionths of a kelvin; nonetheless they are there and represent the point at which matter in the universe first began to organise itself. According to the Big Bang cosmology this picture shows the universe only 700 000 years after it was born and what you are looking at are echoes from the creation!

Cosmology at the frontiers

The ultimate free lunch

Astronomers are constantly making new discoveries and in this final section we will briefly discuss some of the 'hot' issues in cosmology.

Where did all energy in the universe come from? Cosmologists believe that a quantum theory of gravity may provide the answer. Incredible as it may seem, it is possible that all the primordial energy originally came from nothing!

Quantum theory tells us that in a perfect vacuum, which contains neither matter or energy, particle–antiparticle pairs are constantly created or annihilated in a time interval that is too short to observe. These **quantum fluctuations** could lead to a release of energy in a self-creating universe that came into existence entirely spontaneously. Cosmologists have dubbed this apparent energy from nothing 'the ultimate free lunch'.

The cosmological constant

Einstein was the first person to apply general relativity to models of the universe and did so before the Hubble expansion was discovered. In line with current thought at the time, Einstein believed that the universe was unchanging or static, and had existed forever.

In his equations, Einstein found that there was no solution which gave a model of the universe that was static and so introduced a factor called the **cosmological constant** to balance them. When the universe was found not to be static but expanding, the constant was discarded.

However, in the early 1980s, the cosmological constant made something of a 'comeback'. This is because the constant allows a number of solutions to Einstein's equations involving different universe scenarios, and it finds further use in Grand Unified Theories which suggest that the early phase of the evolution of the universe was determined by its value.

Another reason is that the cosmological constant is a useful device in reconciling an apparent discrepancy in the measured ages of the oldest stars in our galaxy, which seem to be greater than the best estimates of H_0 from observation. By adjusting the value of the constant, cosmologists can construct models of the universe that did not expand at such a rapid rate early on, so that the observational evidence is harmonised.

Whether or not the cosmological constant is a necessary 'fudge factor' in cosmological models, the fact remains that its physical meaning is completely unknown.

Space astronomy points the way

At a scientific conference in May 1998, some of the world's leading cosmologists were presented with some startling new evidence. Observations of very distant supernovas as distance indicators suggested that the expansion of the universe, far from being slowed down by gravity, was actually speeding up!

If this was the case then the universe must be filled with some unknown form of matter or energy whose gravity is *repulsive* in nature. This idea, while speculative, indicates that cosmologists are still far from certain about their models of the universe. There is also much to be understood about the nature of matter and the relationship between general relativity and quantum mechanics, and these issues together with others like the formation of galaxies, are matters that still need to be resolved.

Will we ever be able to answer these questions definitively? The way forward undoubtedly lies in obtaining better observational data. Leading the way in this is space astronomy. NASA is currently building a successor to the Hubble Space Telescope called the Next Generation Space Telescope, with superior performance and resolution. Also due for launch in late 2000 is NASA's Microwave Anisotropy Probe (MAP) which will be able to make high resolution measurements of the cosmic microwave background, as will the European Space Agency's *Planck* satellite, scheduled for launch in 2007.

Will the results from these instruments, together with advances in telescope technologies and mathematical insight, provide us with the final answers to the ancient questions of our origin and destiny?

Summary

◆ **Cosmology** today is the study of the universe from the largest to the smallest scales of matter and radiation. Early cosmological models were **geocentric** with the Earth at the centre of the universe and the **Ptolemaic model** was the most successful of these. Copernicus was the first person responsible for changing this view to a Sun-centred or **heliocentric** cosmological model. Sir Isaac Newton realised that gravity held the planets in orbit around the Sun and proposed a model of the universe that was infinite, in which space and time were independent of each other.

◆ Einstein, in his **Special Theory of Relativity**, proposed that the speed of light is constant to all observers and that mass and energy are equivalent. Unlike Newtonian mechanics, space and time become linked together in a **space–time continuum**. In his **General Theory of Relativity** he developed a new theory of gravity in which 'space–time tells matter how to move' and 'matter tells space–time how to curve'.

◆ **Olbers' Paradox** says that the night sky should be uniformly bright if there are an infinite number of stars distributed uniformly in space. It is resolved by the fact that the universe is finite, relatively young and expanding. The **Cosmological Principle** is an assumption made by cosmologists that the universe everywhere is homogeneous and isotropic.

◆ The **Hubble Law** is a relation showing that recession velocity of a distant galaxy is proportional to its distance, and shows that the universe is expanding. The constant of proportionality is called the **Hubble constant** H_0, which changes with time and its reciprocal is the current age of the universe. The precise value of the Hubble constant is uncertain but is thought to be about 75 km s^{-1} per Mpc. The **cosmological red shift** is the red shift observed in the spectra of galaxies due to the universe expanding.

◆ Cosmologists believe that the universe began from a cosmic **singularity** of infinitely high density and temperature that exploded in an event called the **Big Bang**. Strong evidence for a Big Bang is the **cosmic microwave background** which we can observe today. Its uniformity suggests that the universe was homogeneous and smooth in its early history.

◆ Whether the universe will go on expanding or whether it will contract depends on the average distribution of **mass–energy** it contains. There is strong theoretical and observational evidence that suggests that the mass–energy density is close to a critical value, which means that the universe will stop expanding at an infinite time in the future. Geometrically, this corresponds to the shape of the universe being **flat** and having a **density parameter** equal to 1.

◆ Particle physicists have found that matter can be created by **pair production**. The creation of matter early on in the Big Bang involved pair production at very high temperatures. The history of the universe can be divided into four eras of time in which elementary particles, nuclei and the first atoms were successively formed. The formation of light elements in the Big Bang is called **primordial nucleosynthesis** and allows us to make estimates of the abundances of helium in the universe. Radiation and matter were initially locked together but became separate after **decoupling**. From then on matter could start to organise itself via gravity to form the galaxies we see today.

◆ Very early in the Big Bang, the four basic forces or interactions, namely gravity and the electromagnetic, the strong and the weak nuclear forces, were all **unified** in a single **superforce**. As the universe cooled these forces **froze out** to give their different properties as we know them today. **Grand Unified Theories (GUTs)** are used by particle physicists to explain unification, and the Big Bang cosmology offers a way in which these theories can be tested.

◆ Problems with the standard Big Bang model are the **flatness problem** and the **horizon problem**. These can be solved by assuming that the universe expanded very rapidly early in its history using the concept of **inflation**.

◆ It is not possible to model the universe prior to the **Planck time** of 10^{-43} s as there currently exists no physical theory that can describe the universe before that period. The Cosmic Microwave Explorer satellite (COBE) has provided the first evidence of **density fluctuations** in the otherwise smooth microwave background, consistent with the Big Bang model of galaxy formation.

◆ Cosmologists are still far from certain about the definitive model of the universe. It is hoped that space astronomy together with development in mathematical theory linking general relativity and quantum mechanics will provide some of the missing answers.

Questions

1 **a** What is meant by geocentric and heliocentric cosmological models?
  b What is the Cosmological Principle?

2 **a** What is meant by an *inertial frame of reference*? What are the two main postulates of Special Relativity?
  b Explain the terms *length contraction* and *time dilation*.
  c By how much will a metre rule appear to be shortened if it is travelling at i) 0.5c, ii) 0.75c, iii) 0.998c? Comment on your answers. What do they tell you about relative effects?

3 Sub-atomic particles called muons are created by the interaction of cosmic rays with the Earth's upper atmosphere and decay as they reach the ground. Muons travel at speeds close to the speed of light. Using the equations for relativistic mass and time dilation, outline the effects on their mass and half-life due to their very high velocities.

4 What is the Hubble Law? Why is it more accurate to refer to H_0 as the Hubble *parameter* rather than *constant*, and what unit is it usually expressed in? How will H_0 change with time?

5 **a** What is the origin of the microwave background radiation?
  b Explain what is meant by the cosmological red shift. Why does the temperature of the microwave background fall as the universe expands?

6 **a** Give an account of the principal contents of the universe. In your essay you should consider
  i) the relative abundance of the elements and the possible origins of the heavier elements
  ii) electromagnetic radiation and its origins
  iii) the organisation of matter in structures within the universe.
  b Suggest, with reasons, two ways in which the universe might have been different if – all other physical constants remaining the same – the speed of light had been less than it is.

UCLES Cosmology, June 1997

7 Show that the reciprocal of the Hubble constant H_0 has the units of seconds.
  Astronomers are still trying to make accurate measurements of the Hubble constant.
  a A galaxy is moving away from us with radial velocity of 5500 km s^{-1}. Calculate its distance if H_0 is i) 30 km/s/Mpc, ii) 50 km/s/Mpc, iii) 75 km/s/Mpc.
  b How is the age of the universe affected by these differing values of H_0?
  c Suppose the universe stopped expanding and then started contracting. What feature in the spectra of galaxies would enable us to tell that this had happened?

8 a i) With the aid of a labelled diagram, describe the general structure of the Milky Way Galaxy.

ii) On your diagram, mark the appropriate position of the Sun.

b The Earth–Moon distance is approximately 5×10^8 m and the Moon is receding from the Earth at a rate of 0.04 metres per year.

i) Calculate, in Mpc, the distance of the Moon from the Earth.

ii) Write down the Hubble equation for the expansion of the Universe relating galactic speed of recession v to distance d.

iii) One estimate for the Hubble constant H_0 is 60 km s^{-1} Mpc^{-1}. Calculate whether the Moon's recession is faster or slower than that derived from the estimate of H_0.

c Theories for the evolution of the Universe indicate that it may be 'open', 'flat' or 'closed'. The figure is a graph of the variation with time of the size of the Universe to illustrate the three possibilities.

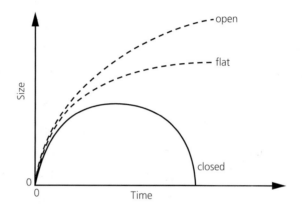

i) Name the physical quantity on which the ultimate fate of the Universe is dependent.

ii) Suggest why it is not yet possible to arrive at an accurate value for the magnitude of this quantity.

OCR Physics, June 1999

Physical, mathematical and astronomical data

Physical data

Physical constants	Symbol	Value
Gravitational constant	G	6.67×10^{-11} N m² kg^{-2}
Acceleration of free fall	g	9.81 m s^{-1} (close to the Earth)
Gravitational field strength	g	9.81 N kg^{-1} (close to the Earth)
Electron charge	e	-1.6×10^{-19} C
Mass of electron	m_e	9.11×10^{-31} kg
Mass of hydrogen atom	m_H	1.67×10^{-27} kg
Unified mass constant	u	1.66×10^{-27} kg
Speed of light in a vacuum	c	3.00×10^{8} m s^{-1}
Molar gas constant	R	8.31 J K^{-1} mol^{-1}
Boltzmann's constant	k	1.38×10^{-23} J K^{-1}
Stefan–Boltzmann constant	σ	5.67×10^{-8} W m^{-2} K^{-4}
Wien's Law constant	$\lambda_{max}T$	2.90×10^{-3} m K
Planck constant	h	6.63×10^{-34} J s
Avogadro constant	N_A	6.02×10^{23} mol^{-1}
Permittivity of free space	ϵ_0	8.85×10^{-12} F m^{-1}
Electronvolt	eV	1 eV $= 1.602 \times 10^{-19}$ J
Temperature	K	K = °C + 273

Mathematical data
Mathematical constants and formulae

$$\pi = 3.1416$$
$$e = 2.7183$$
$$\text{Area of circle} = \pi(\text{radius})^2$$
$$\text{Surface area of a sphere} = 4\pi(\text{radius})^2$$
$$\text{Volume of a sphere} = (4/3)\pi(\text{radius})^3$$
$$\text{Arc length of circle} = \text{radius} \times \text{angle subtended in radians}$$
$$\text{Area of ellipse} = \pi ab \ (a = \text{semi-major axis}; b = \text{semi-minor axis})$$

Angular measure

$$360° = 2\pi \text{ radians}$$
$$1 \text{ radian (rad)} = 57° \, 17" \, 45" = 206\ 264.8"$$
$$1° = 60" = 3600" = 0.017\ 45 \text{ rad}$$

Rules of logarithms

If $10^x = y$ then we say that 'x is the *logarithm* of y to the *base* 10', or in algebraic notation, $\log_{10} y = x$.

For logs of any base a:

Product $\qquad \log_a (AB) = \log_a A + \log_a B$

Quotient $\qquad \log_a (A/B) = \log_a A - \log_a B$

Power $\qquad \log_a x^n = n \log_a x$

Inequalities

Sometimes it is not possible to say exactly what the size of a physical quantity is. Instead we state the limit of its upper or lower value by using inequality symbols:

$\geq$ means is greater than *or* equal to
$\leq$ means is less than *or* equal to
$>$ means is greater than
$<$ means is less than

Significant figures, standard form and rounding

Significant figures are the number of meaningful digits in a numerical quantity. In experimental or observational measurements the accuracy of the method employed will determine the number of significant figures used. Some examples are:

0.0054321	five significant figures
5.4321×10^{-9}	five significant figures
5.1×10^{-6}	two significant figures
5.43	three significant figures

Zeros are not counted at the beginning or end of the number. However, zeros in the middle of a number *are* counted so that, for example, 5.04, 50 400 and 0.0504 all have three significant figures.

In general, when doing calculations the result cannot have more significant figures than the quantity used in the formula having the *least* significant figures. If, for example, one of the quantities in the computation is only accurate to, say, two significant figures (with the rest being known to more than this) then we can only express the result to two significant figures.

In astrophysics we often encounter very large and very small numbers. It is useful to write them in **standard form**, that is, as $a \times 10^n$ where a is a number between 1 and 10, and the index n is an integer (positive or negative whole number, or 0). When multiplying numbers together, the numbers may be multiplied and the indices added. When dividing, the numbers are divided and the indices subtracted. For example:

$$(3.4 \times 10^4) \times (5.6 \times 10^{-3}) = (3.4 \times 5.6) \times 10^{(4-3)} = 19.04 \times 10^1 = 190.4$$
$$(7.5 \times 10^9) \div (4.8 \times 10^{-6}) = (7.5/4.8) \times 10^{(9-(-6))} = 1.6 \times 10^{15}$$

To **round off** numbers we examine the last digit. If it is greater than or equal to 5 we round up, less than 5 we round down, e.g. 34.56 = 34.6 (round up); 976.854 = 976.85 (round down).

Powers of 10

tera (T)	$= 10^{12}$	deci (d)	$= 10^{-1}$
giga (G)	$= 10^9$	centi (c)	$= 10^{-2}$
mega (M)	$= 10^6$	milli (m)	$= 10^{-3}$
kilo (k)	$= 10^3$	micro (µ)	$= 10^{-6}$
hecto (h)	$= 10^2$	nano (n)	$= 10^{-9}$
deca (da)	$= 10$	pico (p)	$= 10^{-12}$

The Greek alphabet

Alpha	A	α	Iota	I	ι	Rho	P	ρ
Beta	B	β	Kappa	K	$\kappa, \varkappa$	Sigma	Σ	σ
Gamma	Γ	γ	Lambda	Λ	λ	Tau	T	τ
Delta	Δ	δ	Mu	M	μ	Upsilon	Y	υ
Epislon	E	ϵ	Nu	N	ν	Phi	Φ	ϕ, φ
Zeta	Z	ζ	Xi	H	ξ	Chi	X	χ
Eta	H	η	Omicron	O	o	Psi	Ψ	ψ
Theta	Θ	θ	Pi	Π	π	Omega	Ω	ω

Astronomical data

Astronomical constants	Symbol	Data
Astronomical unit	AU	$1 \text{ AU} = 1.496 \times 10^{11} \text{ m}$
Parsec	pc	$1 \text{ pc} = 206\ 265 \text{ AU}$ $= 3.26 \text{ ly}$
Light year	ly	$1 \text{ ly} = 6.324 \times 10^4 \text{ AU}$ $= 0.307 \text{ pc}$ $= 9.46 \times 10^{15} \text{ m}$
Mass of Earth	$M_{\oplus}$	$5.98 \times 10^{24} \text{ kg}$
Radius of Earth (equatorial)	$R_{\oplus}$	6378 km
Orbital velocity of Earth	$V_{\oplus}$	30 km s^{-1}
Mass of Sun	$M_{\odot}$	$1.99 \times 10^{30} \text{ kg}$
Radius of Sun	$R_{\odot}$	$6.96 \times 10^5 \text{ km}$
Luminosity of Sun	$L_{\odot}$	$3.90 \times 10^{26} \text{ W}$
Effective temperature of Sun	T_{eff}	5780 K
Mass of Moon	$M_{\mathbb{C}}$	$7.3 \times 10^{22} \text{ kg} = 0.0123 M_{\oplus}$
Radius of Moon	$R_{\mathbb{C}}$	$1738 \text{ km} = 0.273 R_{\oplus}$
Distance of Sun from centre of our galaxy	—	8.5 kpc
Velocity of Sun about galactic centre	V	220 km s^{-1}
Diameter of Milky Way galaxy	—	120 kpc
Mass of Milky Way galaxy	—	$7 \times 10^{11} \ M_{\odot}$

Planetary data

Planet	Mass/ $\times 10^{24}$ kg	Equatorial radius/ km	Average density/ $\times 10^3$ kg m^{-3}	Surface gravity Earth = 1	Orbital period/ days	Escape velocity/ km s^{-1}
Mercury	0.33	2 439	5.4	0.38	87.96	4.2
Venus	4.87	6 052	5.2	0.95	224.70	10.3
Earth	5.98	6 378	5.52	1.00	365.26	11.2
Moon	0.07	1 738	3.34	0.16	27.32 (about Earth)	2.4
Mars	0.64	3 393	3.9	0.39	686.98	5.1
Jupiter	1 900	71 398	1.40	2.74	4 333	61
Saturn	569	60 000	0.69	1.17	10 759	36
Uranus	87	25 559	1.19	0.94	30 685	21
Neptune	103	24 800	1.66	1.15	60 188	24
Pluto	0.01	1 140	0.5(?)	0.03	90 700	2.1

The 20 brightest stars

Star	Constellation	Distance/ parsecs	Apparent magnitude visual	Absolute magnitude visual	Spectral type
Sirius	Canis Major	2.7	−1.5*	+1.4	A1 V
Canopus	Carina	55	−0.7	−3.1	F0 1b
α Centauri	Centaurus	1.3	−0.3*	+4.4	G2 V
Arcturus	Bootes	11	−0.1	−0.3	K2 III
Vega	Lyre	8.1	0.0	+0.5	A0 V
Capella	Auriga	14	0.0*	−0.7	G2 III
Rigel	Orion	250	0.1*	−6.8	B8 1a
Procyon	Canis Minor	3.5	0.3	+2.7	F5 1V–V
Achernar	Eridanus	20	0.5	−1.0	B5 V
β Centauri	Centaurus	90	0.6*	−4.1	B1 III
Altair	Aquila	5.1	0.8	+2.2	A7 1V–V
Betelgeuse	Orion	150	0.8†	−5.5	M2 Iab
Aldebaran	Taurus	16	0.9*	−0.2	K2 III
α Crucis	Crux	120	0.9*	−4.0	B1 IV
Spica	Virgo	80	1.0†	−3.6	B1 V
Antares	Scorpio	120	1.0*†	−4.5	M1 Ib
Pollux	Gemini	12	1.2	+0.8	K0 III
Fomalhaut	Piscis Austrinus	7	1.2	+2.0	A3 V
Deneb	Cygnus	430	1.3	−6.9	A2 Ia
β Crucis	Crux	150	1.3	−4.6	B0.5 IV

* Multiple star system. The apparent magnitude quoted is the combined magnitude of all components.
† Variable star.

The constellations

Name	Genitive form of name	Abbreviation
Andromeda	Andromedae	And
Antlia	Antilae	Ant
Apus	Apodis	Aps
Aquarius	Aquarii	Aqr
Aquila	Aquilae	Aql
Ara	Arae	Ara
Aries	Arietis	Ari
Auriga	Aurigae	Aur
Bootes	Bootis	Boo
Caelum	Caeli	Cae
Camelopardalis	Camelopardalis	Cam
Cancer	Cancri	Cnc
Canes Venatici	Canum Venaticorum	CVn
Canis Major	Canis Majoris	CMa
Canis Minor	Canis Minoris	CMi
Capricornus	Capricorni	Cap
Carina	Carinae	Car
Cassiopeia	Cassiopeiae	Cas
Centaurus	Centauri	Cen
Cepheus	Cephei	Cep
Cetus	Ceti	Cet
Chamaeleon	Chamaelontis	Cha
Circinus	Circini	Cir
Columba	Columbae	Col
Coma Berenices	Comae Berenices	Com
Corona Australis	Coronae Australis	CrA
Corona Borealis	Coronae Borealis	CrB
Corvus	Corvi	Crv
Crater	Crateris	Crt
Crux	Crucis	Cru
Cygnus	Cygni	Cyg

table continues

Name	Genitive form of name	Abbreviation
Delphinus	Delphini	Del
Dorado	Doradus	Dor
Draco	Draconis	Dra
Equuleus	Equulei	Equ
Eridanus	Eridani	Eri
Fornax	Fornacis	For
Gemini	Geminorum	Gem
Grus	Gruis	Gru
Hercules	Herculis	Her
Horologium	Horologii	Hor
Hydra	Hydrae	Hya
Hydrus	Hydri	Hyi
Indus	Indi	Ind
Lacerta	Lacertae	Lac
Leo	Leonis	Leo
Leo Minor	Leonis Minoris	LMi
Lepus	Leporis	Lep
Libra	Librae	Lib
Lupus	Lupi	Lup
Lynx	Lyncis	Lyn
Lyra	Lyrae	Lyr
Mensa	Mensae	Men
Microscopium	Microscopii	Mic
Monoceros	Monocerotis	Mon
Musca	Muscae	Mus
Norma	Normae	Nor
Octans	Octantis	Oct
Ophiuchus	Ophiuchi	Oph
Orion	Orionis	Ori

table continues

223

Name	Genitive form of name	Abbreviation
Pavo	Pavonis	Pav
Pegasus	Pegasi	Peg
Perseus	Persei	Per
Phoenix	Phoenicis	Phe
Pictor	Pictoris	Pic
Pisces	Piscium	Psc
Piscis Austrinus	Piscis Austrini	PsA
Puppis	Puppis	Pup
Pyxis	Pyxidis	Pyx
Reticulum	Reticuli	Ret
Sagitta	Sagittae	Sge
Sagittarius	Sagittarii	Sgr
Scorpius	Scorpii	Sco
Sculptor	Sculptoris	Scl
Scutum	Scuti	Sct
Serpens	Serpentis	Ser
Sextans	Sextantis	Sex
Taurus	Tauri	Tau
Telescopium	Telescopii	Tel
Triangulum	Trianguli	Tri
Triangulum Australe	Trianguli Australis	TrA
Tucana	Tucanae	Tuc
Ursa Major	Ursae Majoris	UMa
Ursa Minor	Ursae Minoris	UMi
Vela	Velorum	Vel
Virgo	Virginis	Vir
Volans	Volantis	Vol
Vulpecula	Vulpeculae	Vul

Answers to numerical questions

Chapter 1

2 a The Earth is in a narrow orbit whereas water is stable: life needs water.

b Density = $\dfrac{\text{mass}}{\text{volume}}$

Average density of Earth

$$= \frac{6.0 \times 10^{24}\ \text{kg}}{\frac{4}{3}\pi(6400 \times 10^3)^3\ \text{m}^3} \approx 5500\ \text{kg m}^{-3}$$

which suggests that the Earth is composed mainly of rock.

3 a Terrestrial and Jovian planets are explained on page 4.

b Using the same method as in **2**:
Average density of Jupiter

$$= \frac{1.9 \times 10^{27}\ \text{kg}}{\frac{4}{3}\pi(71\,400 \times 10^3)^3\ \text{m}^3} \approx 1250\ \text{kg m}^{-3}$$

As well as being much larger and more massive than the Earth, Jupiter is approximately four times less dense.

5 a Look at Figure 1.5 on page 9.

b Time travelled = $\dfrac{\text{distance}}{\text{velocity}} = \dfrac{1.5 \times 10^{11}\ \text{m}}{3 \times 10^8\ \text{m s}^{-1}}$

= 500 s or about 8.3 min

8 a Space astronomy is explained on page 70.

b Time travelled = $\dfrac{\text{distance}}{\text{velocity}}$

One light year = 9.5×10^{12} km, so time taken for *Pioneer 10* to reach *Proxima Centauri* =

$$\frac{4.2 \times 9.5 \times 10^{12}\ \text{km}}{30\ \text{km s}^{-1}} = 1.3 \times 10^{12}\ \text{s}$$

(or since there are 3.15×10^7 s in a year, this is about 41 000 years)

Chapter 2

1 We use the Planck relation (page 23):

$$E_{\text{photon}} = \frac{hc}{\lambda} \text{ and rearranging } \lambda = \frac{hc}{E_{\text{photon}}}$$

Remembering to convert E_{photon} from electronvolts into joules we obtain:

a 0.3 nm (X-ray)
b 300 nm (ultraviolet)
c 30 μm (far infrared)
d 3.1 km (radio)

2 Use $f = \dfrac{E_2 - E_1}{h}$ (page 27)

Transition frequency is

$$\frac{(-1.5\ \text{eV}) - (-3.4\ \text{eV})}{6.63 \times 10^{-34}\ \text{J s}} = \frac{1.9 \times 1.6 \times 10^{-19}\ \text{J}}{6.63 \times 10^{-34}\ \text{J s}}$$

$= 4.6 \times 10^{14}$ Hz (don't forget to convert eV to J!)

The wavelength of the photon is found from:

$$\lambda = \frac{c}{f} = \frac{3.0 \times 10^8}{4.6 \times 10^{14}} = 652.2\ \text{nm}$$

3 We assume that the Sun is perfectly spherical. Surface area of a sphere is given by $4\pi r^2$ so that the Sun's luminosity L is $\sigma \times T^4 \times (4\pi r^2)$ where r is the Sun's radius.

$$L = (5.67 \times 10^{-8}\ \text{W m}^{-2}\ \text{K}^{-4}) \times (5800\ \text{K})^4 \times 4\pi(6.96 \times 10^8\ \text{m})^2$$

$$= 3.9 \times 10^{26}\ \text{W}$$

4 a Stars are good approximations to blackbodies. By Wien's Law, the hotter a blackbody then the shorter the wavelength of peak emission. The Sun is hotter than *Antares* because its wavelength of peak emission (yellow) is shorter than that of *Antares* (red). Look at page 30 for a comprehensive explanation of blackbodies.

b Using Wien's Law

$$\lambda_{\text{max}} = \frac{2.9 \times 10^{-3}}{T}$$

For $T = 2500$ K, $\lambda_{\text{max}} = 1160$ nm
For $T = 35\,000$ K, $\lambda_{\text{max}} = 83$ nm

5 The brightness or light flux received from the Sun is

$$\frac{L_\odot}{4\pi R^2{}_\odot} = \frac{3.90 \times 10^{26}\ \text{W}}{4\pi \times (6.96 \times 10^8)^2}$$

$$= 6.4 \times 10^7\ \text{W m}^{-2}$$

Sunspot = 10% of this = 6.4×10 W m^{-2}. So using Stefan's Law $P = \sigma T^4$

$$T = \left(\frac{P}{\sigma}\right)^{\frac{1}{4}} = \left(\frac{6.4 \times 10^6 \text{ W m}^{-2}}{5.67 \times 10^{-8} \text{ W m}^{-2} \text{ K}^{-4}}\right)^{\frac{1}{4}} = 3300 \text{ K}$$

Using Wien's Law

$$\lambda_{max} = \frac{2.90 \times 10^{-3} \text{ m K}}{3300 \text{ K}} = 878.8 \text{ nm}$$

6 Using the Doppler formula

$$\frac{\lambda - \lambda_0}{\lambda_0} = \frac{v}{c}$$

$\lambda_0 = 600$ nm, $\lambda = 600.8$ nm, so that $\Delta\lambda = +0.8$ nm

$$\therefore v = (3.0 \times 10^8 \text{ m s}^{-1}) \frac{+0.8 \times 10^{-9} \text{ m}}{600 \times 10^{-9} \text{ m}}$$

$$= +400 \text{ km s}^{-1}$$

This is a positive value, so the star is red-shifted and is moving away from us.

7 Since the wavelength difference comes from light at points opposite the Sun's diameter, then $\Delta\lambda = 0.0004$ nm or *one half* of 0.008 nm (i.e. the Doppler shift is ± 0.0004 nm relative to a stationary point).

So $\dfrac{\Delta\lambda}{\lambda} = \dfrac{0.0004 \text{ nm}}{600 \text{ nm}} = 6.7 \times 10^{-6}$

$$\therefore v = c\frac{\Delta\lambda}{\lambda} = (3.0 \times 10^8 \text{ m s}^{-1}) \times (6.7 \times 10^{-6})$$

$$= 2 \text{ km s}^{-1}$$

$$\omega = \frac{v}{r} = \frac{2 \times 10^3 \text{ m s}^{-1}}{7.0 \times 10^8 \text{ m}} = 2.9 \times 10^{-6} \text{ rad s}^{-1}$$

Since 2π radians corresponds to 1 revolution, then

2.9×10^{-6} rad s^{-1} is equivalent to $\dfrac{2\pi}{2.9 \times 10^{-6}} =$

1 revolution in 2.2×10^6 s or about 1 solar revolution every 25.5 days.

8 Using Kepler's 3rd Law for the circular orbit (page 42)

$$\omega = \frac{v}{r} = \sqrt{\frac{GM_{galaxy}}{r^3}} \text{ and rearranging we get}$$

$$M_{galaxy} = r \times \frac{v^2}{G}$$

Putting in the values $r = 2.8 \times 10^4 \times 9.5 \times 10^{15} = 2.7 \times 10^{20}$ m (since 1 light year is about 9.5×10^{12} km) and $v = 2.2 \times 10^5$ m s^{-1}, we get

$$M_{galaxy} =$$
$$(2.7 \times 10^{20} \text{ m}) \times \frac{(2.2 \times 10^5 \text{ m s}^{-1})^2}{6.67 \times 10^{-11} \text{ N m}^2 \text{ kg}^{-2}}$$
$$= 2.0 \times 10^{41} \text{ kg}$$

Since the mass of the Sun ($M_\odot$) is about 2.0×10^{30} kg, the mass of the galaxy is $\approx 10^{11} M_\odot$.

9 a Newton's Law of Universal Gravitation is defined on page 40. It is an inverse square force law.
 b Using the Law of Universal Gravitation ($M_\odot = 2 \times 10^{30}$ kg, 1 AU $= 1.5 \times 10^{11}$ m):

$$F = \frac{\begin{array}{c}(6.67 \times 10^{-11} \text{ N m}^2 \text{ kg}^{-2}) \times (1.9 \times 2 \times 10^{30} \text{ kg}) \\ \times (2.5 \times 2 \times 10^{30} \text{ kg})\end{array}}{(3.2 \times 1.5 \times 10^{11} \text{ m})^2}$$

10 a See page 38 for a definition of Kepler's three Laws of Planetary Motion.
 b Use the 3rd Law in the form $\dfrac{T^2}{r^3} = \dfrac{4\pi^2}{GM_\odot}$

$$\therefore T^2 =$$
$$\frac{4\pi^2 \times (5.20 \times 1.5 \times 10^{11} \text{ m})^3}{(6.67 \times 10^{-11} \text{ N m}^2 \text{ kg}^{-2}) \times (2.0 \times 10^{30} \text{ kg})}$$
$$= 1.4 \times 10^{17} \text{ s}^2$$

so that $T = 3.8 \times 10^8$ s or, since there are about 3.15×10^7 s in 1 year, Jupiter takes nearly 12 years to complete 1 orbit around the Sun.

11 a For a circular orbit, Kepler's 3rd Law is expressed as:

$$\frac{T^2}{r^3} = \frac{4\pi^2}{GM}$$

Using the dimensional notation T for time (seconds), L for length (metres) and M for mass (kilograms), we write Kepler's 3rd Law dimensionally as

LHS $= T^2 L^{-3}$

RHS $= ML^{-3}T^2 \times M^{-1} = T^2 L^{-3} =$ LHS

(Note that G expressed in SI base units is $M^{-1}L^3T^{-2}$.)

 b For a communications satellite to appear to remain in a fixed position above the Earth's surface, its orbital period must be the same as the Earth's period of rotation, i.e. 24 h or 86 400 s. Using Kepler's 3rd Law:

$$\frac{T^2}{r^3} = \frac{4\pi^2}{GM_E}$$

$$r^3 = \frac{T^2 GM_E}{4\pi^2} =$$

$$\frac{(86\,400 \text{ s})^2 \times (6.67 \times 10^{-11} \text{ N m}^2 \text{ kg}^{-2}) \times (6.0 \times 10^{24} \text{ kg})}{4\pi^2}$$

$r^3 = 7.56 \times 10^{22}$ m^3
$\therefore r = 4.23 \times 10^7$ m $= 42\,300$ km

This is the distance from the *centre* of the Earth to the orbital altitude. The radius of the Earth is about 6400 km, so the altitude above the Earth's *surface* is 42 300 km $-$ 6400 km $= 35\,900$ km or nearly 36 000 km.

12 Using Kepler's 3rd Law

$$\frac{T^2}{r^3} = \frac{4\pi^2}{GM_E}$$

and remembering that r = radius of Earth and the altitude above the Earth's surface = 6400 km + 600 km = 7000 km:

$$T^2 = \frac{(7 \times 10^6 \text{ m})^3 \times 4\pi^2}{(6.67 \times 10^{-11} \text{ N m}^2 \text{ kg}^{-2}) \times (6.0 \times 10^{24} \text{ kg})}$$

so that

$T = 5.8 \times 10^3$ s or the HST makes 1 revolution in about 242 days.

To calculate the minimum energy, we use the total energy equation on page 44

$$\text{TE} = \tfrac{1}{2}m_{HST}v^2 - \frac{GM_E m_{HST}}{r}$$

Assuming the HST moves in a perfectly circular orbit, its radius r will be 7000 km so its orbital velocity v is

$$\frac{2\pi r}{T} = \frac{2\pi \times (7 \times 10^6 \text{ m})}{5.8 \times 10^3 \text{ s}} = 7.6 \times 10^3 \text{ m s}^{-1}$$

The total energy is therefore

$$\tfrac{1}{2} \times (11\,000 \text{ kg}) \times (7.6 \times 10^3 \text{ m s}^{-1})^2 -$$

$$\frac{(6.67 \times 10^{-11} \text{ N m}^2 \text{ kg}^{-2}) \times (6 \times 10^{24} \text{ kg}) \times (11\,000 \text{ kg})}{7 \times 10^6 \text{ m}}$$

$$= 4.2 \times 10^7 - 6.3 \times 10^{11}$$

$$= -6.3 \times 10^{11} \text{ J}$$

Note that the negative sign in front of this value tells us that the orbit of the HST is *bounded*, i.e. it has insufficient energy to escape the Earth's gravitational pull.

Chapter 3

1 a The Sun's mass is 2.0×10^{30} kg. You need to look these up but some examples might be:

Mass	Identity
1×10^{-10} (10^{20} kg)	Asteroid or planetary moon
1×10^{-5} (10^{25} kg)	Planet
20 (10^{31} kg)	High mass star

b

Distance (or diameter)	Identity
1.5 light seconds = 4.5×10^8 m	Approx. Earth–Moon distance
6×10^4 ly ≈ 5.7×10^{20} m	Approx. radius of Milky Way galaxy
1×10^7 ly ≈ 9.5×10^{22} m	Average distance between galaxies

2 Since the light spreads out as an inverse square law, the observed luminosity of the Sun as seen from Pluto would be

$$\frac{3.90 \times 10^{26} \text{ W}}{(39.44 \text{ AU})^2} = 2.5 \times 10^{23} \text{ W}$$

The Sun would appear dimmer by a factor of $39.44^2 \approx 1560$ AU.

3 b i) Using $d = \frac{1}{p''}$ pc

we find

$$d = \frac{1}{0.133} \text{ pc} = 7.52 \text{ pc}$$

ii) 1 pc = 206 264.81 AU, so
7.52 pc = 7.52 × 206 264.81 = 1 550 000 AU
iii) 1 pc = 3.262 ly, so
7.52 pc = 7.52 × 3.262 = 24.5 ly

4 a See page 55. The Sun has a positive absolute magnitude because absolute magnitudes of all stars are defined at a standard distance of 10 pc from the Earth. The Sun is much closer than this, so its apparent magnitude is negative whereas its absolute magnitude is defined as if it was 10 pc distant.

b We use $m - M = 5 \log d - 5$
with $m = -1.5$ and $M = +4.77$ so that
$(-1.5) - (+4.77) = 5 \log d - 5$
$-6.27 = 5 \log d - 5$

$$\therefore \log d = \frac{-1.27}{5} = -0.254$$

$$d = 10^{-0.254} = 0.56 \text{ pc}$$

5 a $m - M = 5 \log d - 5$ so that
$M = m - 5 \log d + 5$
$d = 16$ ly $= 16 \div 3.262 = 4.9$ pc and
$M = (+0.77) - 5 \times \log(4.9) + 5 = 2.3$
Altair has an absolute magnitude of $+2.3$.

b $m - M = 5 \log d - 5$ so that
$d = 10^{(m-M+5)/5}$
$d = 10^{((+1.35)-(-0.3)+5)/5} = 10^{1.33} = 21$ pc

c We use $M = 4.77 - 2.5 \log(L/L_\odot)$ so that

$$M = 4.77 - 2.5 \log\left(\frac{100 L_\odot}{L_\odot}\right) = -0.23$$

Capella's absolute magnitude is -0.23.
$d = 10^{(m-M+5)/5}$ so that
$d = 10^{((+0.08)-(-0.23)+5)/5} = 10^{1.062} = 11.5$ pc

6 a LGP of eye $= \frac{1}{4}\pi(0.5 \text{ cm})^2$
LGP of telescope $= \frac{1}{4}\pi(20 \text{ cm})^2$

$$\frac{\text{LGP}_{\text{telescope}}}{\text{LGP}_{\text{eye}}} = \frac{(20 \text{ cm})^2}{(0.5 \text{ cm})^2} = 1600$$

The telescope has 1600 times the light gathering power of the human eye.

$$\text{Magnification} = \frac{\text{focal length of objective}}{\text{focal length of eyepiece}}$$

Focal length of objective $= 240$ cm
i) $(2400 \text{ mm})/(6 \text{ mm}) = 400$
ii) $(2400 \text{ mm})/(8 \text{ mm}) = 300$
iii) $(2400 \text{ mm})/(24 \text{ mm}) = 100$

b Compared with the human eye the LGP of the HST is

$$\frac{(240 \text{ cm})^2}{(0.5 \text{ cm})^2} = 230\,400$$

The LGP of the HST is over 200 000 times that of the human eye.

Image resolution $\theta_{\min} = 1.22\dfrac{\lambda}{a}$

For HST $a = 2.4$ m

At $\lambda = 500$ nm, $\theta_{\min} = 2.5 \times 10^{-7}$ rad
At $\lambda = 200$ nm, $\theta_{\min} = 1.0 \times 10^{-7}$ rad
At $\lambda = 2000$ nm, $\theta_{\min} = 1.0 \times 10^{-6}$ rad

10 a Dish size

b Resolving power $=$

$$\frac{1.22\lambda}{L} = \frac{1.22 \times (1.5 \times 10^2 \text{ m})}{(30 \times 10^3 \text{ m})}$$

$= 6.1 \times 10^{-3}$ rad

$L =$ diameter of Earth $= 1.28 \times 10^7$ m

$\therefore$ Resolving power $=$

$$\frac{1.22 \times (1.5 \times 10^2 \text{ m})}{(1.28 \times 10^7 \text{ m})}$$

$= 1.4 \times 10^{-5}$ rad

11 a Annual parallax and method is defined on page 52. The method assumes that the Earth's orbit is perfectly circular.
Let parallax angle be p.

Then $\tan p = \dfrac{1.50 \times 10^{11} \text{ m}}{8.7 \times 9.46 \times 10^{15} \text{ m}}$

$= 1.8 \times 10^{-6}$ degrees
$= (\pi/180) \times 1.8 \times 10^{-6} = 3.1 \times 10^{-8}$ rads
$= 3.1 \times 10^{-8} \times 206\,264.81$ arc seconds
$= 6.6 \times 10^{-3}$ arc seconds

b The luminosity of a star is the number of joules of energy it radiates per second, i.e. its *power*. Using the inverse square law:

$$\text{Intensity of } \textit{Sirius A} = \frac{8.17 \times 10^{27} \text{ W}}{(8.7 \times 9.46 \times 10^{15} \text{ m})^2}$$

$= 1.2 \times 10^{-6}$ W m^{-2} (notice units of luminosity are *watts*, those of intensity *watts per square metre*)

Chapter 4

2 Using $E_{\text{photon}} = \dfrac{hc}{\lambda}$

$$E_{\text{photon}} = \frac{(6.63 \times 10^{-34} \text{ J s}) \times (3.0 \times 10^8 \text{ m s}^{-1})}{588 \times 10^{-9} \text{ m}}$$

$= 3.38 \times 10^{-19}$ J (or 2.1 eV)

The dark line corresponds to absorption of light when an electron in the helium atom is raised from the level corresponding to -5.80×10^{-19} J to that of -2.43×10^{-19} J, as the energy difference between these levels is very close to the photon energy at 588 nm. This suggests the presence of helium in the solar atmosphere.

3 a Use the Rydberg formula $\dfrac{1}{\lambda} = R\left(\dfrac{1}{2^2} - \dfrac{1}{n^2}\right)$

$n = 3, 4, 5$. The first one is calculated as

$n = 3$ $\dfrac{1}{\lambda} = 1.0974 \times 10^7\left(\dfrac{1}{2^2} - \dfrac{1}{3^2}\right)$

$\therefore \lambda = 656.1$ nm

The others are
$n = 4$	$\lambda = 486.0$ nm	$n = 9$	$\lambda = 383.4$ nm
$n = 5$	$\lambda = 433.9$ nm	$n = 10$	$\lambda = 379.7$ nm
$n = 6$	$\lambda = 410.1$ nm	$n = 11$	$\lambda = 377.0$ nm
$n = 7$	$\lambda = 396.9$ nm	$n = 12$	$\lambda = 374.9$ nm
$n = 8$	$\lambda = 388.8$ nm		

b If we put $n = \infty$ into the Rydberg formula, then

$$\frac{1}{\lambda} = 1.0974 \times 10^7\left(\frac{1}{2^2} - 0\right) \text{ with } \lambda = 364.5 \text{ nm}$$

as the limit of wavelength.

5 *Antares, Arcturus, Alpha Centauri A, Fomalhaut, Sirius, Spica.*
Alpha Centauri A is most like the Sun.

Chapter 5

2 From its colour using photometry.

We assume that the light spreads out from *Capella* as an inverse square law. If the intensity of the radiation received from *Capella* is 1.2×10^{-8} W m^{-2} and it is 4.3×10^{17} m distant, then the luminosity at *Capella* must be $(1.2 \times 10^{-8}$ W m$^{-2}) \times$ $(4.3 \times 10^{17}$ m$)^2 = 2.2 \times 10^{27}$ W. Since $L = 4\pi r^2 \sigma T^4$

$$r = \sqrt{\frac{L}{4\pi\sigma T^4}}$$

$$= \sqrt{\frac{2.2 \times 10^{27} \text{ W}}{4\pi \times (5.7 \times 10^{-8} \text{ W m}^{-2} \text{ K}^{-4}) \times (5200 \text{ K})^4}}$$

$$= 2.0 \times 10^9 \text{ m}$$

The radius of *Capella* is 2.0×10^9 m or about $2.9 R_\odot$.

3 From the brightness–time graph, the period T of the binary pair is 18 days. This equals $18 \times 3600 \times 24 = 1\,555\,200$ s.

Angular speed $\omega = (2\pi)/T = \dfrac{2\pi}{1\,555\,200 \text{ s}}$

$$= 4.04 \times 10^{-6} \text{ rad}$$

The brightness–time graph shows that the stars are partially eclipsing binaries. The graphs representing stars X and Y are in anti-phase with one star moving towards the observer while the other is moving away. The wavelength shift of star X is less because its radial velocity is less than star Y. Their angular speed is the same because they both have the same period.

$$v_X = c \times \frac{\Delta\lambda}{\lambda_0}$$

$$= \frac{(3.0 \times 10^8 \text{ m s}^{-1}) \times (0.120 \times 10^{-9} \text{ m})}{447 \times 10^{-9} \text{ m}}$$

$$= 80.5 \text{ km s}^{-1}$$

$$r_1 = \frac{v_X T}{2\pi}$$

$$= \frac{(80.5 \times 10^3 \text{ m s}^{-1}) \times (1\,555\,200 \text{ s})}{2\pi}$$

$$= 2.0 \times 10^{10} \text{ m}$$

$$v_Y = c \times \frac{\Delta\lambda}{\lambda_0}$$

$$= \frac{(3.0 \times 10^8 \text{ m s}^{-1}) \times (0.150 \times 10^{-9} \text{ m})}{447 \times 10^{-9} \text{ m}}$$

$$= 100.7 \text{ km s}^{-1}$$

$$r_2 = \frac{v_Y T}{2\pi}$$

$$= \frac{(100.7 \times 10^3 \text{ m s}^{-1}) \times (1\,555\,200 \text{ s})}{2\pi}$$

$$= 2.5 \times 10^{10} \text{ m}$$

$r = r_1 + r_2 = 4.5 \times 10^{10}$ m (or 0.30 AU)

Chapter 6

2 $\lambda_{max}T = 2.90 \times 10^{-3}$ m K

$$\therefore \lambda_{max} = \frac{2.90 \times 10^{-3}}{30 \text{ K}} = 96.7 \text{ μm}$$

This is in the infrared region of the electromagnetic spectrum.

3 There are 0.33 hydrogen atoms per 10^6 m^3 or 3.3×10^5 atoms m^{-3}. This represents a mass of $(3.3 \times 10^5) \times (1.7 \times 10^{-27}) = 5.6 \times 10^{-22}$ kg m^{-3}.

$\therefore 10^6$ m^3 has a mass of 5.6×10^{-16} kg attributed to hydrogen atoms, and this mass forms 99% of the interstellar matter. Since there is one dust particle per 10^6 m^3, the total mass of interstellar matter in this volume is $(5.6 \times 10^{-16}$ kg$)/0.99 =$ 5.7×10^{-16} kg, and one dust particle has a mass of 1% of the total $= 5.7 \times 10^{-18}$ kg. If each dust particle has a density of 3000 kg m^{-3} and is spherical, then

$$3000 \text{ kg m}^{-3} = \frac{5.7 \times 10^{-18} \text{ kg}}{\frac{4}{3}\pi r^3 \text{ m}^3}$$

$$r = \left(\frac{5.7 \times 10^{-18} \text{ kg}}{\frac{4}{3}\pi \times (3000 \text{ kg m}^{-3})}\right)^{\frac{1}{3}} = 7.7 \times 10^{-8} \text{ m}$$

4 1 pc $= 3.1 \times 10^{16}$ m

Volume of nebula $= \frac{4}{3}\pi \times (3.1 \times 10^{16}$ m$)^3$

$= 1.2 \times 10^{50}$ m^3

The number of hydrogen atoms in this volume is therefore $(10^9$ atoms m$^{-3}) \times (1.2 \times 10^{50}$ m$^3)$

$= 1.2 \times 10^{59}$.

Total energy to ionise the nebula is $(1.2 \times 10^{59}) \times 13.6$ eV $= 1.6 \times 10^{60}$ eV or 2.6×10^{41} J.

5 a Mass of hydrogen $= 1.67 \times 10^{-27}$ kg so that density of cloud $= (10^{18} \times 1.67 \times 10^{-27})$

$= 1.67 \times 10^{-9}$ kg m^{-3}

$$t_{ff} = \frac{2.11 \times 10^{-3}}{(1.67 \times 10^{-9} \text{ kg m}^{-3})^{1/2}} = 52 \text{ years}$$

b $t_{ff} = \dfrac{6.65 \times 10^4}{(1400 \text{ kg m}^{-3})^{1/2}} \approx 1800$ s

6 a $m = (-7.1) + 5\log(276) - 5 = +0.1$

b $m = (-7.1) + 5\log(276) - 5 +$ $(0.002 \times 276) = +0.7$

The star appears dimmer by a factor of 7 in magnitude due to interstellar absorption (remember the more positive the magnitude is, the dimmer the star appears).

Chapter 7

2 Using $\Delta E = \Delta m \times c^2$

$$\Delta m = \frac{E}{c^2} = \frac{8.4 \times 10^{19}\,J}{(3.0 \times 10^8\,m\,s^{-1})^2} = 933\,kg$$

4 Mass of two deuterons $= 2 \times 2.015 = 4.030$ u
Mass of isotope of helium plus neutron
$= 3.017 + 1.009 = 4.026$ u
Mass defect $= 4.030 - 4.026 = 0.004$ u
Equivalent energy $= 0.004 \times 931$ MeV $= 3.7$ MeV
$(6.0 \times 10^{-13}\,J)$

6 $E_{photon} = \dfrac{hc}{\lambda}$

$$\lambda = \frac{(6.63 \times 10^{-34}\,J\,s) \times (3.0 \times 10^8\,m\,s^{-1})}{5.49 \times 10^6 \times 1.6 \times 10^{-19}\,J}$$

$= 2.3 \times 10^{-13}$ m

9 See pages 133, 136 and 143.
To overcome the electrostatic repulsion of the nuclei. Higher temperatures are needed in CNO cycle because repulsion between the nuclei is greater.
It takes 4 H nuclei to produce 1 He nucleus, so mass defect $= 4 \times (1.0078\,u - 4.0026\,u) = 0.0286$ u.
Using the mass–energy relation $\Delta E = \Delta mc^2$, the equivalent energy is 4.33×10^{-12} J.
If 3×10^{14} moles s^{-1} of H are converted into He then the number of joules per second is:

$\frac{1}{4} \times (3 \times 10^{14}) \times (6.02 \times 10^{23}) \times (4.33 \times 10^{-12})$
$= 2.0 \times 10^{26}$ J s^{-1}

10 Equivalent mass of 4×10^{26} J is

$$\Delta m = \frac{4 \times 10^{26}\,J}{c^2} = 4.4 \times 10^9\,kg$$

Therefore mass loss $= 4.4 \times 10^9$ kg s^{-1}.
Assuming the Sun uses all its hydrogen, it will shine for

$$\frac{2 \times 10^{30}\,kg}{4.4 \times 10^9\,kg\,s^{-1}} = 4.5 \times 10^{20}\,s$$

(or 1.4×10^{13} years)

(The Sun will not actually shine for this long as it will undergo dramatic change before all its hydrogen is consumed.)

Chapter 8

4 $T_{ms} = 10^{10} \dfrac{M_{star}}{M_\odot} \times \dfrac{L_\odot}{L_{star}}$

For $25M_\odot$, $T_{ms} = 10^{10} \dfrac{25M_\odot}{M_\odot} \times \dfrac{L_\odot}{80\,000L_\odot}$

$= 3 \times 10^6$ years

Using the same method: $15M_\odot$, 15×10^6 years; $3M_\odot$, 500×10^6 years; $1.5M_\odot$, 3000×10^6 years; $1.0M_\odot$, $10\,000 \times 10^6$ years; $0.75M_\odot$, $15\,000 \times 10^6$ years; $0.5M_\odot$, $200\,000 \times 10^6$ years.

5 a See page 151.
 b Mass defect $= (3 \times 4.0026\,u) - 12.0000$ u
$= 0.0078$ u
$\Delta E = \Delta m \times c^2 = 0.0078$ u $\times 931$ MeV
$= 7.26$ MeV
 c See page 154.

Chapter 9

2 a Use $g = \dfrac{GM}{r^2}$ (page 41).

$$g = \frac{(6.67 \times 10^{-11}\,N\,m^2\,kg^{-2}) \times (1.4 \times 10^{30}\,kg)}{(1.5 \times 10^6\,m)^2}$$

$= 4.2 \times 10^7$ m s^{-2} (or N kg^{-1})

 b For 1 kg mass use GPE $= -\dfrac{GM}{r}$ (page 44).

GPE $=$

$-\dfrac{(6.67 \times 10^{-11}\,N\,m^2\,kg^{-2}) \times (1.4 \times 10^{30}\,kg)}{1.5 \times 10^6\,m}$

$= -6.3 \times 10^{13}$ J kg^{-1}

 c Use $v_{esc} = \sqrt{\dfrac{2GM}{r}} =$

$$\sqrt{\frac{2 \times (6.67 \times 10^{-11}\,N\,m^2\,kg^{-2}) \times (1.4 \times 10^{30}\,kg)}{1.5 \times 10^6\,m}}$$

$= 1.1 \times 10^7$ m s^{-1}

This is about 4% the speed of light.

3 KE $= \frac{1}{2}mv^2 = \frac{1}{2} \times (2 \times (2 \times 10^{30})) \times$ $(5 \times 10^6\,m\,s^{-1})^2 = 5 \times 10^{43}$ J

4 Use conservation of angular momentum (page 170).

$$\left(\frac{\omega_{neut}}{\omega_{star}}\right) = \left(\frac{R_{star}}{R_{neut}}\right)^2$$

$$\therefore \omega_{neut} = \omega_{star} \times \left(\frac{R_{star}}{R_{neut}}\right)^2$$

$$\omega_{neut} = \frac{2\pi}{2.6 \times 10^6 \text{ s}} \times \left(\frac{15 \times 10^8 \text{ m}}{10 \times 10^3 \text{ m}}\right)^2$$

$$= 5.4 \times 10^4 \text{ rad s}^{-1} \text{ (period} = 0.12 \text{ ms)}$$

5 Use $g = \frac{GM}{r^2}$ so that

$$M = \frac{gr^2}{G}$$

$$= \frac{(9.81 \times 10^{11} \text{ N kg}^{-1}) \times (8 \times 10^3 \text{ m})^2}{6.67 \times 10^{-11} \text{ N m}^2 \text{ kg}^{-2}}$$

$$= 9.4 \times 10^{29} \text{ kg (or } 0.47M_\odot)$$

Emission from accreting matter (PE → KE, converted to heat during accretion).

6 Use $R_{Sch} = \frac{2GM}{c^2}$ (page 174).

Putting in the values for $M = M_\oplus$, $M_\odot$ and M_{galaxy} ($\approx 10^{11} M_\odot$), we obtain:
a 8.9 mm
b 3.0 km
c 3.0×10^{14} m (0.03 ly)

The Schwarzschild radius is proportional to the mass of an object. More massive objects have larger R_{Sch}. (Remember that R_{Sch} is the critical radius that an object must have in order to be dense enough so that its gravitational field can stop light escaping and become a black hole.)

8 a & b See page 44.
c i) $M_\odot = \rho_\odot \times V_\odot$
$= (1.4 \times 10^3 \text{ kg m}^{-3}) \times (1.4 \times 10^{27} \text{ m}^{-3})$
$= 2.0 \times 10^{30} \text{ kg}$

ii) $v_{esc} = \sqrt{\frac{2GM_\odot}{R_\odot}}$

Putting in the values of $M_\odot$, $R_\odot$ and G gives $6.2 \times 10^5 \text{ m s}^{-1}$.

d $v_{esc} = \sqrt{\frac{2GM_\odot}{R_\odot}}$ with $R = 12$ km

$= 1.5 \times 10^8 \text{ m s}^{-1} (\frac{1}{2}c)$

e $R = \frac{2GM_\odot}{(v_{esc})^2}$

Putting $v_{esc} = 3.0 \times 10^8 \text{ m s}^{-1}$ gives $R = 3$ km.

f $KE = \frac{1}{2}mv^2 = \frac{1}{2} \times (2.0 \times 10^{-3} \text{ kg}) \times (1.5 \times 10^8 \text{ m s}^{-1})^2 = 2.3 \times 10^{13}$ J

g From the kinetic theory of gases we can write $\frac{1}{2}mc^2 = \frac{3}{2}kT$

$$\therefore T = \frac{\frac{1}{2}mc^2}{\frac{3}{2}k} = \frac{2.3 \times 10^{13} \text{ J}}{1.5 \times (1.38 \times 10^{-23} \text{ J K}^{-1})}$$

$$= 1.1 \times 10^{36} \text{ K}$$

h i) $E_{photon} = \frac{hc}{\lambda} = kT$

$$\therefore \lambda = \frac{hc}{kT} =$$

$$\frac{(6.6 \times 10^{-34} \text{ J s}) \times (3.0 \times 10^8 \text{ m s}^{-1})}{(1.38 \times 10^{-23} \text{ J K}^{-1}) \times (10^{12} \text{ K})}$$

$$= 1.4 \times 10^{-14} \text{ m}$$

ii) Gamma ray region.

Chapter 10

2 c Use $I = I_0 \sqrt{\left(1 - \frac{v^2}{c^2}\right)}$ (page 188)

   i) $v = 0.5c$, 13.4%
   ii) $v = 0.75c$, 33.9%
   iii) $v = 0.998c$, 93.7%
   Relativistic effects get more pronounced at speeds close to c.

7 Use Hubble Law $v = H_0 \times r$ (page 195)

So $\frac{1}{H_0} = \frac{r}{v} =$ dimensionally: $\frac{L}{LT^{-1}} = T$ (seconds)

   a i) 183 Mpc
   ii) 110 Mpc
   iii) 73 Mpc

   b Age of universe $= \frac{1}{H_0}$

   ∴ a larger $H_0 = $ a younger universe.

   c The red shifts of the galaxies would change to blue shifts.

8 a i) & ii) see Chapter 1.

   b i) 1 Mpc $= 3.1 \times 10^{22}$ m

   Earth–Moon distance $= \frac{5.0 \times 10^8 \text{ m}}{3.1 \times 10^{22} \text{ m}}$

   $= 1.6 \times 10^{-14}$ Mpc

   ii) $v = H_0 \times d$

   iii) Using $H_0 = 60$ km s^{-1} Mpc^{-1}

   $v = (60 \text{ km s}^{-1} \text{ Mpc}^{-1}) \times (1.6 \times 10^{-14} \text{ Mpc})$

   $= 9.7 \times 10^{-13}$ km s^{-1}

   Recession velocity of the moon

   $= \frac{0.04 \times 10^{-3} \text{ km}}{(365 \times 24 \times 3600 \text{ s})} = 1.3 \times 10^{-12}$ km s^{-1}

   ∴ the Moon's recession is *faster* than that derived from the estimate of H_0.

   c i) The physical quantity is the *density parameter* Ω_0 (see page 201).

   ii) At the moment it is difficult to obtain accurate measurements of the total mass–energy in the universe.

Bibliography

Internet

An abridged version of this book is reproduced on the University of Surrey Physics website at **http://www.ph.surrey.ac.uk/astrophysics/index.html**. The site also contains many links to relevant material on astronomy, astrophysics and cosmology.

Books

The New Solar System, edited by J. Kelly Beatty, C. Collins Peterson and A. Chaikin. Sky Publishing Corporation 1999 (fourth edition 1999)
An excellent account of the many new discoveries made in the Solar System by space probes and what they mean. Contains many photographs and illustrations.

Journeys to the Ends of the Universe: A guided tour of the beginnings and endings of planets, stars, galaxies and the universe, C.R. Kitchin. Institute of Physics Publishing (1990)
A series of essays which takes the reader on a guided tour of the new and exciting concepts which are now shaping our views of the universe.

Telescopes and Techniques, C.R. Kitchin. Springer (1995)
Intended for first-year astronomy degree students, this book nonetheless contains much that can be understood by A-level students, including non-technical descriptions of many techniques used in astronomical and astrophysical measurements.

Black Holes, Wormholes and Time Machines, J.S. Al-Khalili. Institute of Physics Publishing (1999)
A primer on the physics of black holes and wormholes and the theoretical possibility of time travel. This is a non-technical book aimed at a teenage audience that explores the strange properties of space and time near black holes.

Astronomy before the Telescope, edited by Christopher Walker. British Museum Press (1996)
A collection of essays about astronomy in ancient times and before the invention of the telescope. Early models of the universe are discussed, including the achievements of European astronomers from the prehistoric to medieval periods and their links with the ancient cultures of Egypt and Mesopotamia.

A Short History of the Universe, J. Silk. Scientific American Library (1994)
A highly readable book explaining current developments in cosmology and their relation to particle physics. There is no mathematics involved and all the various concepts are explained in a lucid way.

Particle Physics, Christopher Bishop. John Murray (2000)
A companion to this book in the *Advanced Physics Readers* series. Includes a section on cosmology.

A Brief History of Time – from the Big Bang to Black Holes, S.W. Hawking. Bantam Press (1988)
A book by one of the foremost thinkers in cosmology and astrophysics today, Stephen Hawking's excellent and accessible account of the creation of the universe and its ultimate fate includes speculations on the nature of time and what might have happened before 10^{-43} s. Despite the complexity of the subject matter, there are many things that A-level students will be familiar with, having first read this book on astrophysics.

New Astronomer: The Practical Guide to the Skills and Techniques of Skywatching, Carole Stott. Dorling Kindersley (1999)
Astrophysics is not about reading books! You need to get under the night sky and see the universe for yourself! This excellent book, by a former astronomer at the Royal Observatory Greenwich, tells you how to get started and what to see. Contains much useful advice on buying and using telescopes.

Periodicals

Sky and Telescope
Astronomy Now
Astronomy
These both keep the reader up to date with the latest developments in astronomy and astrophysics and contain many articles of general interest.

Telescope suppliers

Broadhurst Clarkson & Fuller, Telescope House, 63 Farringdon Road, London EC1M 3JB
A long established company that stocks a wide range of telescopes and astronomical accessories. They supply schools and colleges and are able to give practical and unbiased advice in choosing a telescope.

Societies

British Astronomical Association, Burlington House, Piccadilly, London W1V 9AG

For information on how you can help reduce light pollution and make the universe easier to see at night, contact:
British Astronomical Association Campaign for Dark Skies, 38 The Vineries, Colehill, Wimborne, Dorset BH21 2PX

Index

Index

MK spectral classification
system 96
molecular clouds 118–19
molecular spectra 93
moment of inertia 171
Moon, Full, magnitude of 54
moons 4
Morgan, W.W. 95
Morley, Edward W. 187, 188
muons 190

negative curvature 203
neon burning 152, 156
Neptune 4
neutrino 136
neutrino oscillation 142
neutron capture 153
neutron-degeneracy pressure
162, 180
neutron stars 13, 161–2, 180
Newton, Sir Isaac 1, 213
Laws of Motion 37, 40, 47,
186
Universal Law of Gravitation
1, 37, 40–2, 47, 103, 104,
186
Newtonian focus 64
Next Generation Space
Telescope 212
noise 79
non-inertia 191
northern lights see aurora
borealis
novas 14, 167, 180
nuclear atom 130–5
nuclear binding energy 130,
131, 144
nuclear fission 133, 144
nuclear fusion 133, 144
Nuclear Test Ban Treaty (1963)
71
nucleon 26
number 26

nucleosynthesis 151
nucleus 26

objective 62
Olbers, Heinrich Wilhelm 193
Olbers' paradox 193, 213
Oort, Jan H. 6
Oort Cloud 6
opaque universe 138
optical activity 78
optical telescopes 62–6
limitations 66
in use 65
optical window 25
orbits and energy 44
Orion 102, 116, 119, 149
Orion Nebula 116, 117, 119,
123
outer core, Earth's 4
oxygen burning 152–3, 156

pair production 204–5, 214
parabola 44
parallax 1, 50–3, 82
parallax angle 51, 59
parsec (pc) 14, 50–3, 82
particle–antiparticle pair 204–5
particle physics 203
Paschen series 28
Pauli Exclusion Principle 159
peak wavelength 30
Penrose, Sir Roger 176
penumbra 10
Penzias, Arno 199
period–luminosity (P–L) relation
60–2, 82
photoelectric effect 72
photographic emulsion 72
photometers 77
photometry 77
photomultiplier tubes (PMTs)
72–3

photon
absorption of 27
definition 23, 47
emission of 27
energy of 23–5, 47
temperature of 205
photosphere 10, 33
pixels 74
Planck curve 31, 47
Planck relation 23, 27, 47
Planck satellite 212
Planck time 209–10, 214
Planck's constant 23
planet 4, 21
planetary nebula 151, 156, 180
plasma, solar 12
plasma state 92
Pleiades 120
Pluto 4, 38
Pogson, Norman 55
Pogson's Law 55–6, 57, 60, 82,
101, 197
polarimeters 78
Polaris 60, 184
polarisation 120, 121, 127
polarisation elements 78
population I stars 108
population II stars 108
positive curvature 203
positron 136
post-helium burning 152–3
post-hydrogen burning 143,
144
post-main sequence 146, 156
potential barrier 135
potential energy (PE) 44
precession 187
pre-main sequence 146
pre-main sequence star 125,
127, 146, 156
prime focus 64
primordial nucleosynthesis 206,
214
Principle of Equivalence 192